THE SEA AROUND US
LABORATORY MANUAL

Mark Pagani
Michael A. Arthur
Albert L. Guber

Department of Geosciences
Pennsylvania State University
University Park, Pa 16802

KENDALL/HUNT PUBLISHING COMPANY
4050 Westmark Drive   Dubuque, Iowa 52002

Copyright © 1994 by Mark Pagani, Michael A. Arthur, and Albert L. Guber

ISBN 0-7872-0063-8

All rights reserved. No part of this publication may be reproduced, stored in a retrieval system, or transmitted, in any form or by any means, electronic, mechanical, photocopying, recording, or otherwise, without the prior written permission of the copyright owner.

Printed in the United States of America
10   9   8   7   6   5   4   3   2   1

Contents

1. The Geographic Grid — 1
2. Bathymetry — 9
3. Sea-Floor Spreading — 19
4. Waves — 31
5. The Beach Environment — 43
6. Barrier Islands — 57
7. Salinity — 65
8. Ocean Stratification — 75
9. Ocean Nutrient Dynamics — 89
10. Plankton — 103
11. Deep-Sea Vent Communities — 113
12. Salt Marsh Environment — 117

Answers to Additional Questions — 123

Color Photographs:
- Barrier Island — 125-135
- North Atlantic Phytoplankton Productivity — 136-139
- Global Biosphere — 140

THE GEOGRAPHIC GRID

The shape and size of the earth

The earth is spherical but bulges somewhat around the equator and is compressed near the poles. In cross-section it is shaped like an ellipse. For our purposes, however, we will consider the earth as a perfect sphere with a diameter of approximately **7,927 statute (English) miles.** The evidence for the earth's spherical form comes from several observations that can be made without instruments or calculations. Circumnavigation of the globe and the curved shadow of the earth as it is projected on the moon during a lunar eclipse are two early observations which led explorers and scientists to believe that the earth was spherical (though one could have argued that for both cases the earth was cylindrical or disk shaped). A third proof can be found at sea. As passing ships recede farther into the distance, they appear to sink slowly below the horizon. Through a telescope, the bottom of a distant ship disappears below sea level first, followed by the decks and masts. The sea surface, and thus the earth's surface, must be curved. On the open ocean this phenomenon occurs in any direction and thus supports a spheroid rather than a cylindrical shape. Photographs of the earth from the moon or from a satellite are some modern proofs of its spheroidal shape.

The ancient Greeks believed the earth had a spherical shape as far back as 540 B.C. The earth's circumference, however, was not determined until about 200 B.C. Eratosthenes, a librarian at Alexandria, observed that on the day of the summer solstice (and *only* on that day) the sun's noon rays penetrated to the bottom of a vertical well that was located at Syene, Egypt (close to the Tropic of Cancer, 23.5°N). That is to say, when the sun was at its zenith, its rays where perpendicular to the surface of the earth. He also noted that in Alexandria (925 km north of Syene - originally measured in units called *stades*), on the same day, at the same time, the rays of the sun were not perpendicular to the earth's surface, but arrived at an angle of 7° 12' with respect to a vertical line. Assuming that the sun's rays are parallel when they reach the earth and that the earth is a sphere, Eratosthenes was able to calculate the circumference of the earth with surprising accuracy.

In Figure 1, **Point (C)** is at the center of the earth and the sun is shown at its zenith over Syene (**B**). At Alexandria (**A**), 925 km to the north, the sun (not at its zenith) makes an angle (a) of 7° 12' (decimal equivalent = 7.2°) with a ver-

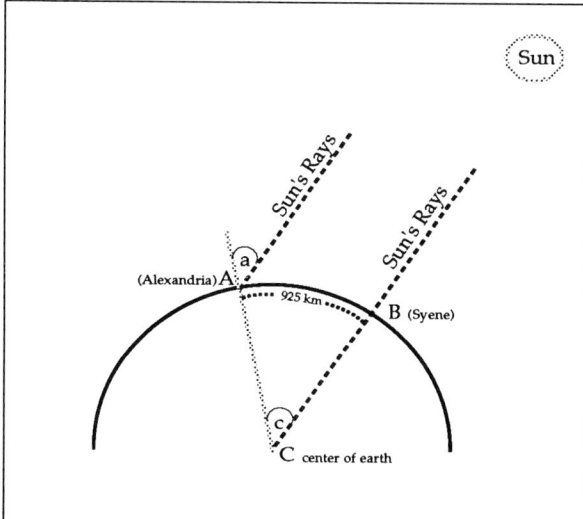

FIGURE 1-1. *Eratosthenes calculation of the circumference of the earth.*

tical line. If the sun's rays are parallel, **angle (a)** must equal **angle (c)**. Angle (c) represents a fraction of the 360 degrees that are inherent to a circle and **arc A-B** is equivalent to 5,000 stades. The calculation at this point would be:

$$\frac{7.2°}{360°} = \frac{925 \text{ km}}{x}$$

where x is the earth's circumference. Solving for x yields:

$$x = 46{,}250 \text{ km}$$

Eratosthenes calculated the earth's circumference to be 250,000 stades or 46,250 km (about 26,660 statute miles). **The true value is 40,000 km (24,903 statute miles).** His error resulted from the fact that Syene was slightly north of the Tropic of Cancer, Alexandria was not exactly north of Syene, and there was a minor error in the distance measured between Syene and Alexandria. His measurement, however, was the best determination of the earth's circumference for more than 1,000 years.

The Coordinates

For the purpose of locating any point on the earth, a series of intersecting lines, known as the **geographic grid**, has been developed. Lines running east-west, parallel to the equator, are called **parallels** (lines of latitude). Lines running north-south, connecting the poles, are called **meridians** (lines of longitude). Latitude and longitude are really angles measured in degrees of arc along a circle, with the center of the earth as the center of the circle. Each **degree of arc** can be subdivided into **60 minutes of arc** (1°= 60') and each minute into **60 seconds of arc** (1'= 60"). When referring to a position on the earth's surface, latitude, by convention, precedes longitude, and both latitude and longitude are recorded with their hemisphere notation (north-south, east-west). For example: 25°17′ 30″ N, 30°25′ 00″ E. This is read, 25 degrees 17 minutes 30 seconds north latitude, 30 degrees 25 minutes 0 seconds east longitude.

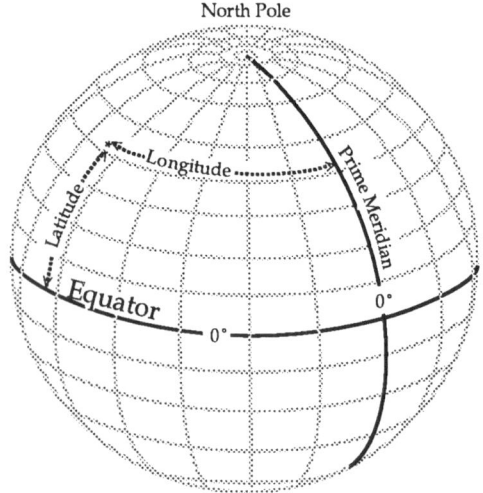

FIGURE 1-2. *The geographic grid.*

Longitude

Meridians are arcuate lines that run north-south along the surface of the earth. The intersection of meridians around the earth's surface inscribe a circle. These circles are all the same size and intersect at both poles. Because they intersect, they are not parallel to each other and are farthest apart at the equator. Lines of longitude range from 0° to 180° in both east and west directions (because a sphere has a circumference of 360°). 0° Longitude is known as the **prime meridian**, and, by definition, passes through the Royal Observatory at Greenwich, England. East of this line is known as east longitude. West of the prime meridian is collectively known as west

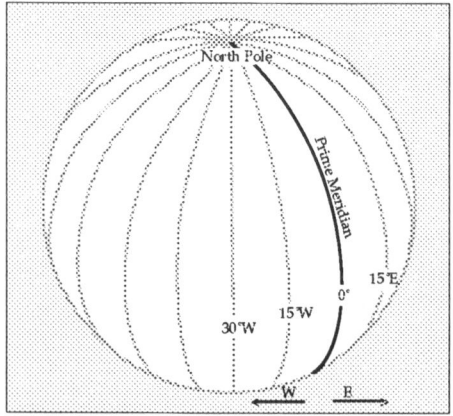

Figure 1-3. Meridians.

longitude. Because longitude lines converge at the poles, *the actual length of one degree of longitude will depend upon where you measure it*. For example, at the equator, the distance of one degree of longitude will be the circumference of the earth divided by 360°;

$$\frac{24,903}{360} = 69.18 \text{ statute miles}$$

As you approach the poles, the distance of one degree of longitude decreases.

Latitude

Lines of latitude are parallel circular lines that run east-west. The equator divides the globe

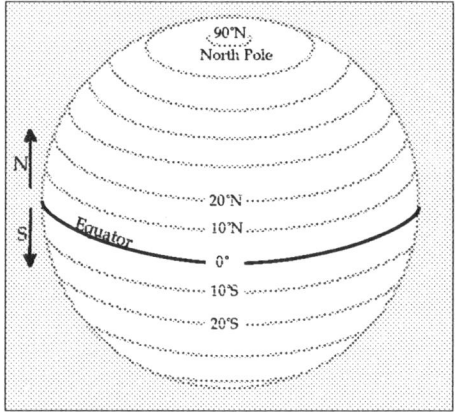

FIGURE 1-4. *Parallels.*

into the *northern* and *southern* hemispheres and is designated 0° latitude. Parallels range from 0° to 90° (at the poles) north and south of the equator. North of the equator is north latitude, south of the equator is south latitude. The circles formed by the intersection of the earth's surface with the planes of latitude become progressively smaller as you approach the poles, with the equator as the largest circle. Unlike meridians, parallels are equidistantly spaced, so that *one degree of latitude is equivalent to the same distance no matter where it is measured*. At the equator, one degree of latitude corresponds to one degree of longitude (approximately 69 statute miles). Therefore, *one degree of latitude is equivalent to 69 statute miles anywhere on earth*.

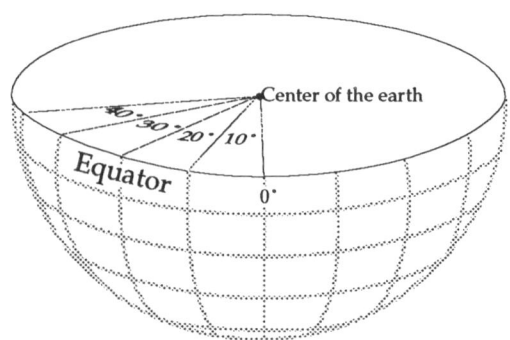

FIGURE 1-5. *Both longitude and latitude are angles measured from the center of the earth.*

Determination of Latitude

Two methods, prior to modern technology, have been used to determine latitude. One uses a distant star and the other, the sun. In the northern hemisphere, the star most commonly used is the North Star (**Polaris**). Polaris is positioned almost directly above the north pole and is visible throughout the northern hemisphere. At the north pole, Polaris would appear directly overhead, but at the equator, the star appears just at the horizon. That is to say, its elevation is 0° at the equator and 90° at the north pole (which also correspond to degrees of latitude). If one travels north from the equator, Polaris would appear to rise from the horizon. The **angle of inclination** from the horizon to the North Star corresponds to your latitude.

Other stars can be used to determine latitude, but because they are not positioned over the polar zenith, a correction must be used. The southern hemisphere has no bright star positioned over it's polar zenith, and so there is no way to determine latitude directly.

Solar latitude determination is more complex and beyond the scope of this lab.

Determination of Longitude

In 24 hours, the earth spins on it's axis 360 degrees. *If you divide 360° by this 24 hour period, you will calculate the number of degrees the earth spins in one hour.* This is equivalent to **15°/hr**. Obviously, longitude and time are closely related. World time zones are generally based on standard meridians, which are multiples of 15°. For purposes of navigation, longitude can be determined by comparing local time with that of Greenwich, England (0° longitude). Every hour difference corresponds approximately to 15°of longitude. You should also remember that it will be *later east* of some point, and *earlier west* of that point. For example: at local noon (your time) you determine that it is 9:00 AM Greenwich time, i.e. 3 hours later than Greenwich, England (0° longitude). 3 hours is equivalent to 45° longitude, and because your time is later than Greenwich time, you must be east of it. Therefore your position is longitude 45° east.

An interesting situation exists with longitude and time. A position 180° east of the of the prime meridian corresponds to a time difference of 12 hours later in the day. A position 180° west of the prime meridian corresponds to 12 hours earlier in the day. At the 180° meridian there exists a 24 hour discrepancy just east and west of that line. In other words, if you are crossing the Pacific Ocean, a 24 hour correction is made as you pass the 180° meridian. For example, if you were on a plane traveling *westward* from Seattle to some point in Asia, and prior to crossing the 180th meridian it was 2:00 pm Monday, it would be 2:00 pm Tuesday after you crossed the 180° meridian (*time moves ahead by one day*). Traveling east across this line moves your apparent time one day earlier.

Because of these interesting conditions, the 180th meridian was named the **International Date Line**. As with all time zones, political boundaries have forced the the International Date Line to deviate somewhat to accommodate land areas and island groups that require schedules having the same calendar day.

Distance and Speed

As you know, the term 'mile' is an expression of distance. Since there are different types of miles, however, it is more accurate to refer to the English mile as a **statute mile** (5,280 ft). At sea, distance is measured in **nautical miles.** *One nautical mile is equivalent to one minute of latitude.* If there are 60 minutes in one degree, there must be *60 nautical miles in one degree.* A nautical mile is, therefore, slightly larger than a statute mile [1 nautical mile = 1.15 statute miles]. (Remember, one degree of latitude is also equivalent to 69 statute miles.)

Speed on land is measured in miles/hour or kilometers/hour, while at sea the unit of speed is the **knot**. *A knot is one nautical mile/hour.* Notice that the definition of knot includes the term "per hour", thus it is redundant and incorrect to say knots/hour.

Distance to the Horizon

On the sea, the distance to the horizon is given by the equation:

$$D = \sqrt{1.5 H}$$

Where: H is the height of the observer's eye above sea level in feet. D is the distance to the horizon in nautical miles. (A nautical mile is equivalent to 1.15 statute miles).*

Along with the observer's height, visibility conditions and wave height will also play a part in determining the observed distance to the horizon.

* NOTE: This equation does not fit the rules of "dimensional analysis," but is expressed this way for convenience of calculation.

Definitions

Equator. 0° latitude. Separates the globe into northern and southern hemispheres.

Circumference. The distance around a circle.

Diameter. A line segment that passes through the center of a circle, and whose end points lie on the circle.

International Date Line. A jagged line, approximately equivalent to the 180th meridian (approximately located in the Pacific Ocean), where a date change occurs.

Knot. The unit of speed used at sea. Equivalent to 1 nautical mile/hour.

Latitude. Angular distance north or south of the equator measured from 0° at the equator to 90° at the poles.

Longitude. Angular distance east or west of the prime meridian measured from 0°(Greenwich, England) to 180°.

Meridians. Lines of longitude.

Nautical mile. The unit of distance used at sea. Equivalent to 6,080 feet, 1.15 statute miles, or 1.85 kilometers.

Parallels. Lines of latitude.

Prime meridian. 0° longitude. Passes through Greenwich, England.

Radius. Half the distance of a diameter.

Statute mile. A unit of distance used on land. Equivalent to 5,280 feet, .87 nautical miles, or 1.6 kilometers.

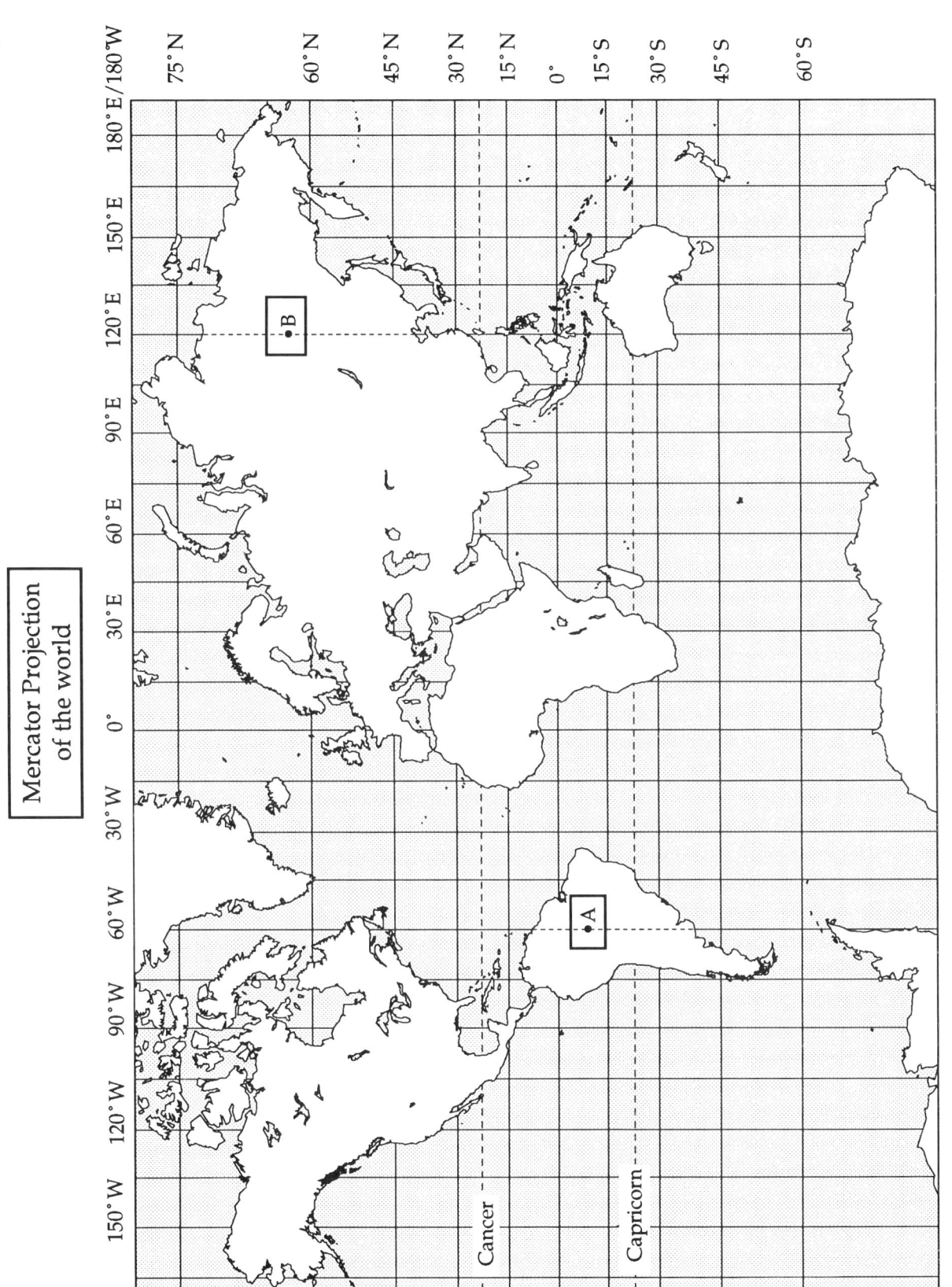

Table 1

Pi (π) = **3.1416**
Circle: **360° of arc**
1 degree of arc = **60 minutes of arc**
1 minute of arc = **60 seconds of arc**

Circumference = π x **Diameter**
Velocity = distance/time
Radius = Diameter/2

Statute mile = **5,280 feet**
Nautical mile = **6,080 feet**
1 nautical mile = **1.15 statute miles**
Knot = **1 nautical mile/hr**
1 knot = **1.85 km/hr**

1° of latitude = **60 nautical miles = 69 statute miles**
Earth's diameter = **7,927 statute miles**

Additional Questions for Review

1. Calculate the equatorial circumference of the earth in statute miles. Convert your answer to nautical miles.

2. What is the speed of any fixed point on the equator moving through space?   (v=distance/time)

3. As you move away from the equator (north or south) is your speed: the same, lower, faster? Why?

4. If your position is 32°40' 00" N, 40°25' 00" E, what would be the angle of inclination to the star Polaris?

5. If the earth rotates 15° in one hour, how many degrees of arc does it move in one minute (of time)? In one second (of time)?

6. If you are standing 20°10' 15" N, 23°45'12" W and you move 5°20' to the west, what will be your new position?

7. (Refer to the Mercator Projection provided). When it is exactly noon in Greenwich (prime meridian), what time is it at point A? At point B?

8. A boat is traveling 20 knots. Convert this to statute miles/hour.

9. It is 2 pm where you are (30° E longitude). At another place, point B, it is 9:00 am. What is point B's longitude? What is its latitude?

10. Point A is at latitude 45°25' N and longitude 90°20' W. Point B is at 25°25' N, 73°45' W.

    (a) How much farther to the south is point B relative to point A, in nautical miles and in statute miles?

    (b) What is the difference in longitude between points A and B?

    (c) What is the difference in time between the two points?

BATHYMETRY

Bathymetry is defined as the measuring and charting of the topography of the sea floor. An intimate knowledge of bathymetry is important to a navigator operating a vessel in shallow water. Early efforts in the determination of depth measurements (known as **soundings**) were accomplished by dropping a weighted line, which was divided into fathoms (1 fathom = 6 feet) by markers, from a vessel. The ordinary length of a line was on the order of 40 fathoms, and therefore soundings were commonly restricted to shallow water. Deep-sea soundings, using longer line, took several hours to complete, and were characteristically inaccurate because of stretch in the long lines being deployed as well as drift causing the line to move away from a vertical orientation.

In 1922, the U.S. Navy developed a device, known as the **echo sounder**, that revolutionized sea floor imagery. The echo sounder (or fathometer) records the time a sound pulse travels from the vessel to the sea floor and back again. Knowing that sound travels 1470 meters per second in sea water on average, one can easily determine depth. The echo sounder is much faster and more accurate than dropping a line, and can be used with the ship under way. As the ship travels, the echo sounder emits sound pulses, records the travel time of the *reflected* sound pulses, converts the time to distance, and then draws this information on a graph known as a **strip-chart recorder**. An **echogram** is the resulting display on the strip chart. Precision Depth Recorders are routinely used on any oceanographic vessel. Their ubiquitous use over the last several years has developed a moderately detailed image of the sea floor.

The Echo Sounder

The sound pulse of an echo sounder is generated by a device known as a *transponder*, which also serves as a receiver for the returning signal. The sound generated is in the form of a broad cone (generally with an apex of 15°). The first return signal received by the transponder will be that which was reflected from the closest reflecting surface. This may be a point directly beneath the vessel (which is what the ship is intending to record) or it may be some other high point off to the side of the ship. An isolated high point off to the side may produce what is known as a **false bottom** on the strip-chart recorder. The false bottom will usually appear long before the ship is actually over the rise. As the vessel approaches the peak of the submarine rise (the peak directly beneath the ship) the two traces (the actual bottom and the false bottom) will converge. False bottoms tend to appear as "hats" on the tops of high points. A sharp depression in the sea floor

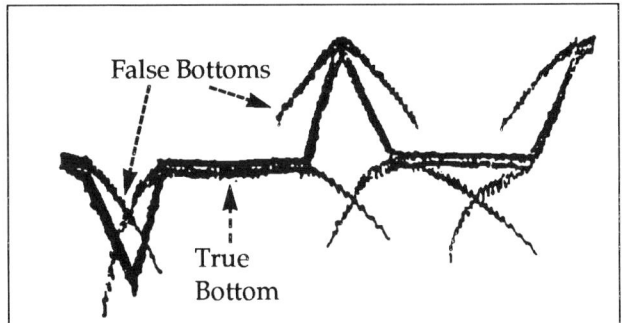

FIGURE 2-2. *False bottoms.*

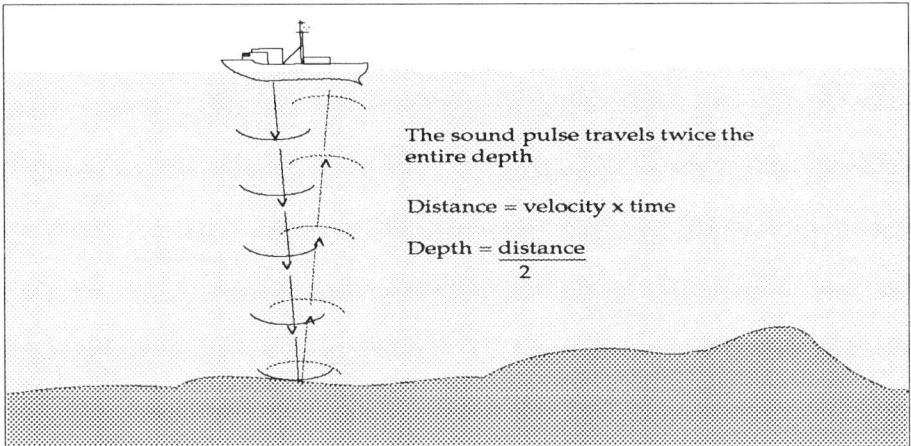

FIGURE 2-1. *Echo sounder.*

will also generate a false bottom and tend to make the **relief** (the differences in elevation between two points) appear less than it actually is.

The sound pulse will reflect off of any surface whose density is different than that of sea water. Thus, the sound generated onboard travels downward to the sea floor and is reflected back towards the ship, at which point it is received by the transponder. The reflected sound pulse, however, does not necessarily end its journey. Much of the sound pulse can continue up to the surface of the sea where it is then *reflected back down* to the sea floor and subsequently back to the ship. In this case, the sound pulse makes two round trips to the bottom. The sea floor that is recorded in this case is known as a **multiple**. As many as three or four multiples may be recorded depending on the strength of the signal. Each multiple is *displaced downward from the previous one by a distance equal to the water depth.*

Relief

Relief is the difference between *maximum* and *minimum* depths. **Local relief** refers to locally adjacent hills and valleys, where **total relief** refers to the difference between the highest and lowest points on the entire profile. Relief can be determined by subtracting the water depth at the base of a feature with the water depth at its peak.

Vertical exaggeration

Because depth profiles can cover very large horizontal distances, it is common to exaggerate the vertical axis in order to see the sea floor topography. Without this distortion, the sea floor would appear as virtually flat on most depth profiles. Determination of vertical exaggeration is fairly simple. For example: If the profile's *horizontal scale* indicates that 1 cm is equivalent to 1000 meters, and the *vertical scale* indicates that 1 cm is equivalent to 50 meters, then vertical exaggeration would equal: **Vertical exaggeration = 20x**

$$\frac{1000 \text{ meters}}{50 \text{ meters}} = 20x$$

The value for any vertical exaggeration will always be *greater than one*. The greater the vertical exaggeration, the steeper the slope of a feature will appear.

Contour lines

After many soundings have been recorded in a particular area, they are eventually marked on a base map. From the base map, contour lines can be drawn. **Contours** are lines which connect points of equal depth, elevation, or concentration. The shape and spacing of contours express the three dimensional sea floor (or any surface) on a two dimensional page. The general rule is that close, crowded contour lines represent a steep slope, and widely space lines represent a gentle slope. In developing a contour map from plotted soundings there are several rules which might be helpful:

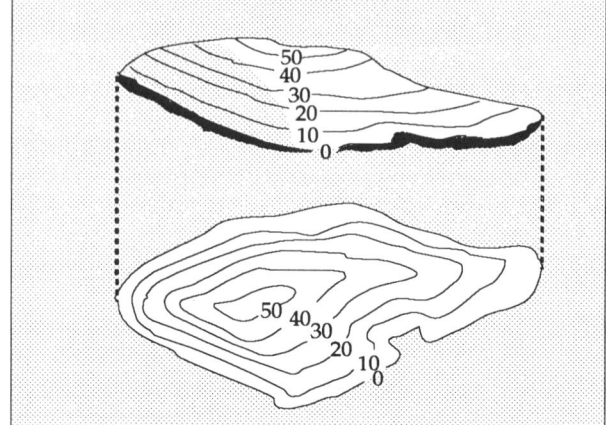

FIGURE 2-3. Contour lines. An example of a hill-like feature projected on a flat surface.

1. Contour lines never cross
2. Contour lines do not end. (They will eventually join somewhere off the map)
3. Always label contour lines *on the map*.
4. When crossing a deeply incised valley or canyon, the contour line will have a V-shape, with the point of the V directed upstream.
5. Close spacing indicates steep slopes. Wide spacing indicates gentle slopes.
6. Sketch first in pencil.

Definitions

Bathymetry. The measuring and charting of the topography of the sea floor.

Contour line. A map line that connects points of equal elevation.

Contour interval. The difference in elevation between adjacent contours.

Multiple. A seismic wave that has more than one reflection. On a depth profile, it is plotted as an incorrect depth for the sea floor. The depth of each multiple is equal to the depth of the water between the vessel and sea floor.

Relief. The difference in elevation between two points. The points are not necessarily adjacent.

Sounding. A measurement of water depth from the surface of the sea to the sea floor.

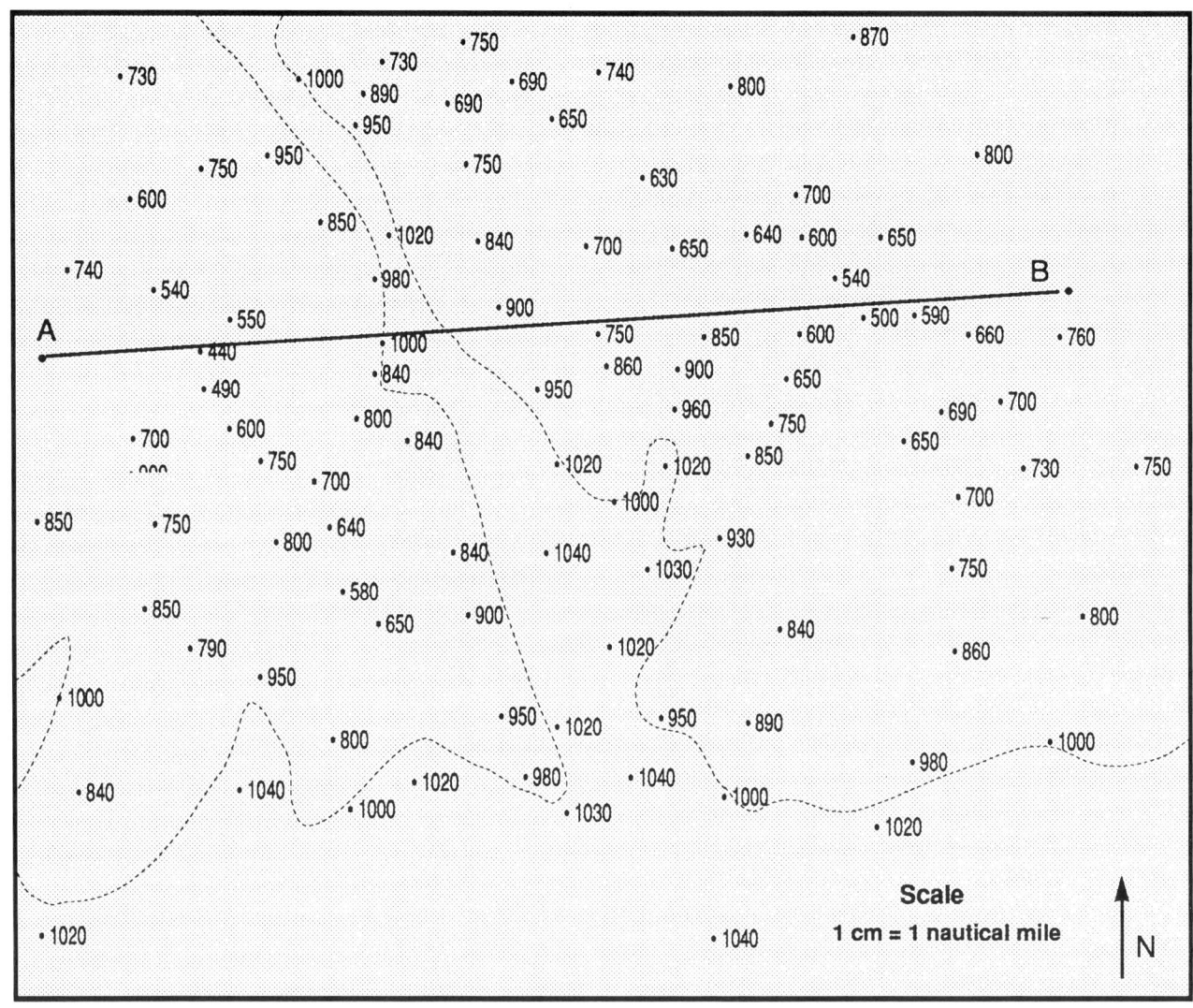

Base Map. (Dotted line represents 1000 meter contour)

Exercise

1) On the base map above contour (in pencil) the soundings on using a 100 meter contour interval.

2) Describe the general appearance of the features drawn on the base map.

3a) Construct a profile between the points A and B using Grid 1 on the following page. To do this, place the top of the grid along line A-B and mark off the contour lines that cross the top line of the grid. Each contour represents a depth, thus each tick mark should be moved downward to its appropriate depth. Once all the depths have been identified, connect the points with a smooth line.

b) What is the vertical exaggeration of your profile?

c) Redraw your profile without vertical exaggeration on Grid 2. Do this by dropping vertical lines from Grid 1 down to Grid 2. Mark off the appropriate depths and connect the points with a smooth line.

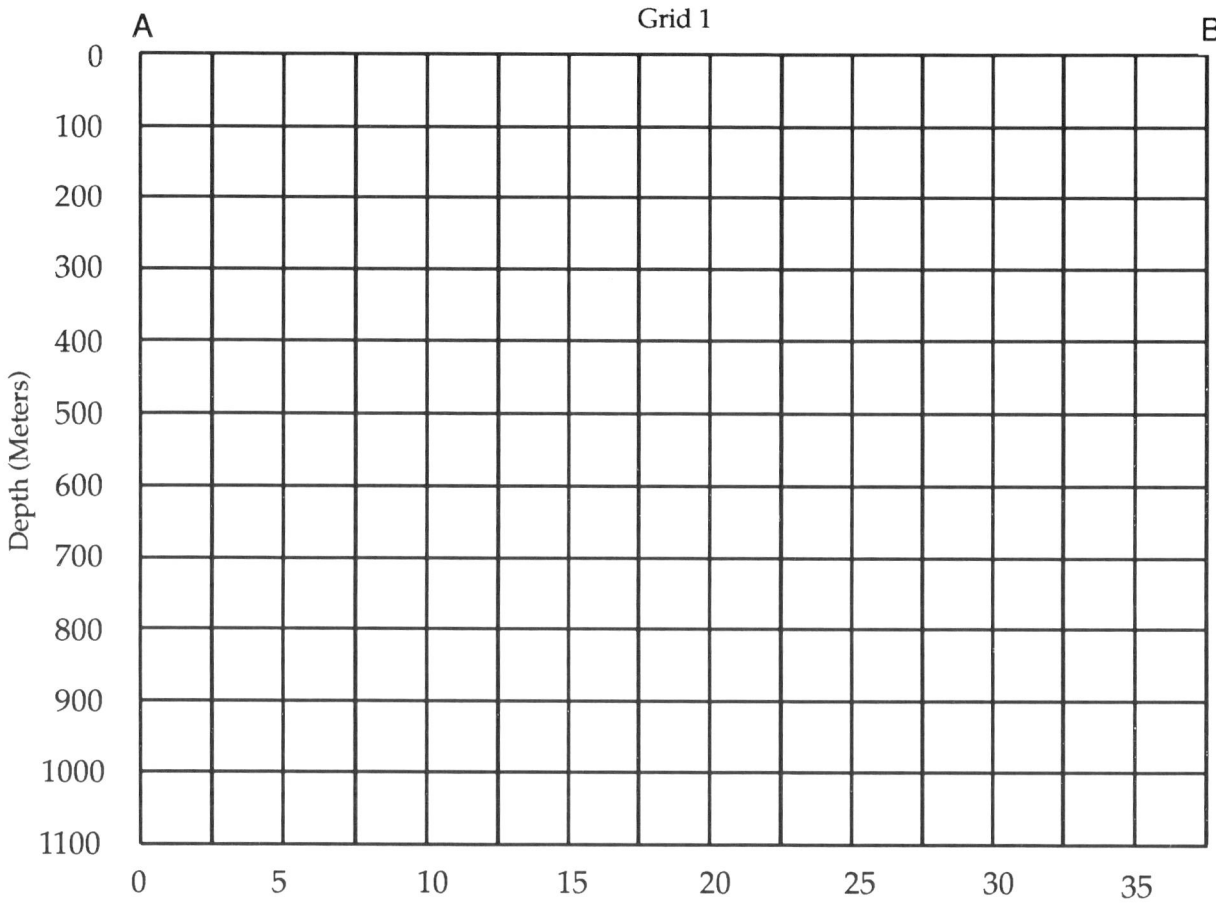
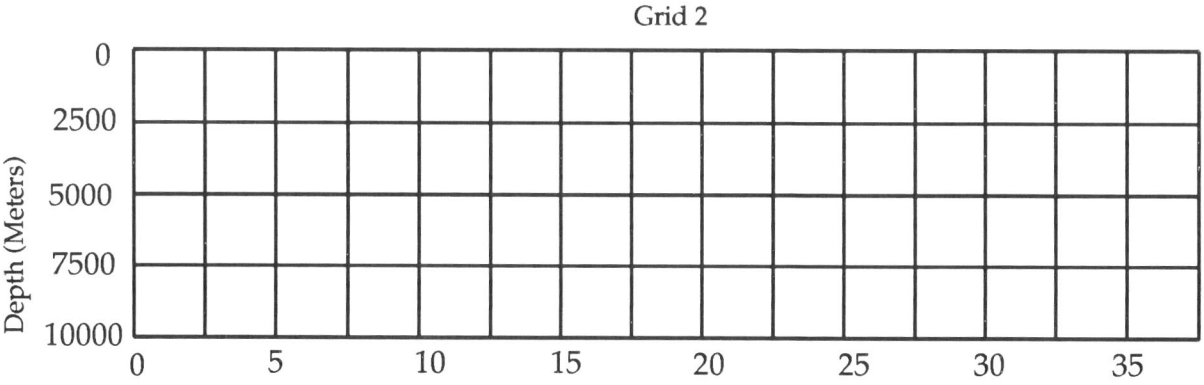

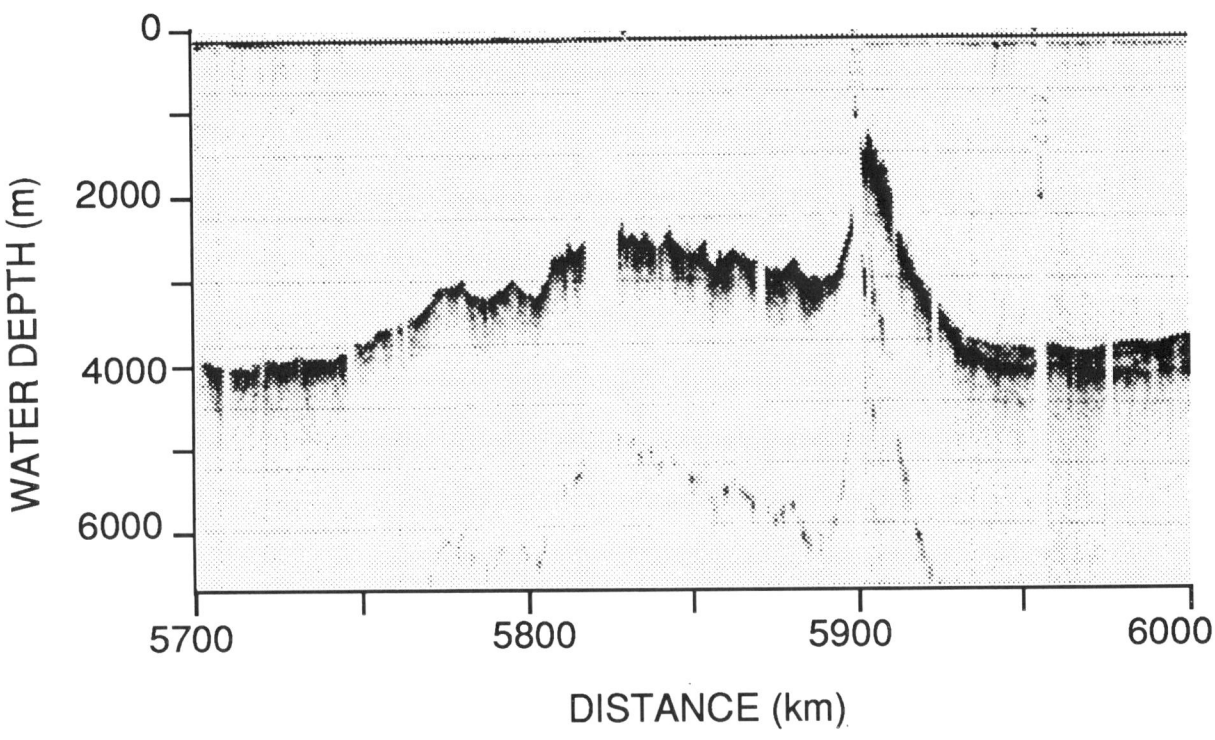

Exercise

1) From the echogram provided determine vertical exaggeration. Show all your work.

2a) What is the feature recorded faintly below the shallower bottom echo?

b) Describe how this feature is formed.

3) What is the total relief of the profile?

Additional Questions for Review

1. If it takes 5 seconds for a sound pulse to leave and return to a ship, how deep is the sea floor below? (Sound travels 1470 meters/second or 4800 feet/second).

   In meters:

   In feet:

2. If the vertical scale of an echogram is 1 cm = 50 meters and the horizontal scale is 1 cm = 10 km, what is the vertical exaggeration?

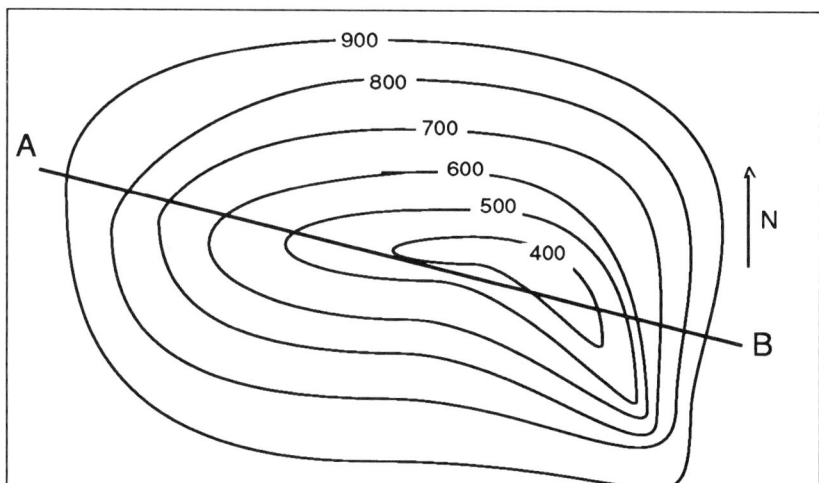

3(a). Describe the feature represented by the contours.

   (b). Is the eastern slope steeper or gentler than the western slope?

   (c). What is the total relief along profile A - B?

4. How long will it take a sound wave to return to a ship if the sea floor below the vessel is at a depth of 14,400 feet?

SEA-FLOOR SPREADING

Plate Tectonics

The earth is differentiated by density and composition into three layers: the crust, mantle and core (which has a liquid outer layer and a solid inner core). The earth's outer layers can be subdivided further into two zones: the **lithosphere** and the **asthenosphere**. The lithosphere is the rigid, outermost shell approximately 100 km thick. It contains all of the crust, including the ocean floors and the continents, and the upper portion of the mantle. This outer shell is mobile and rides on the underlying, partially molten asthenosphere, which extends down approximately 250 km from the base of the lithosphere. The lithosphere is broken into six large, rigid **crustal plates** and several smaller ones. Plate tectonics is based on the motion of these plates as they ride on the asthenosphere as distinct units. *As these crustal plates move apart, new oceanic crust is produced by intrusion of basalt.* This process occurs at **mid-ocean ridges** known as **divergent zones** or lithospheric plate sinks (is subducted) below the other plate, and eventually the subducted plate is incorporated back into the mantle. Mountain building, volcanism and earthquakes accompany the zones associated with both accretion and subduction of oceanic crust, with the most frequent and violent earthquakes along subduction zones.

In addition to the divergence and convergence of plates, plate segments can also move laterally or slip past each other. Along these *fracture zones* are a type of fault known as **transform faults**. Mid-oceanic ridges are broken by many of these faults at approximately right angles to the ridge crest. These faults allow for the motion of rigid lithospheric plates about the spherical earth. A commonly known transform fault is the San Andreas fault of California.

In summary, boundaries around an individual plate can be characterized as either *divergent zones, subduction zones, or transform faults*. These plate boundaries, in turn, are characterized by active seismic activity and volcanism. Some

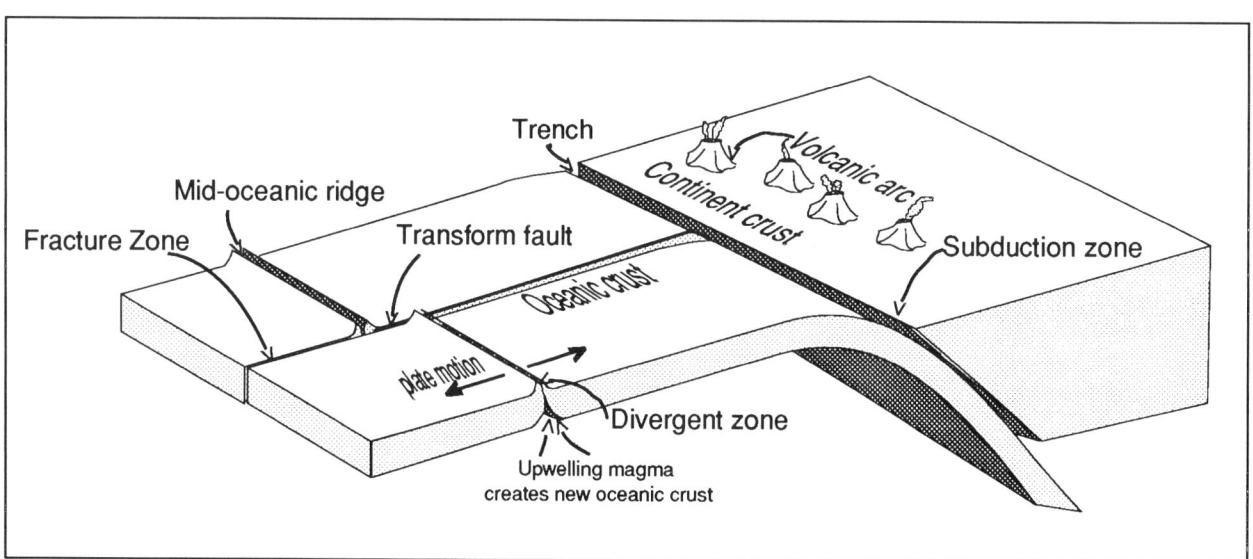

FIGURE 3-1. *Plate tectonics. Plates diverge along the axis of mid-oceanic ridges. Spreading plates are replaced by upwelling magma, which cools to form oceanic crust. Removal or subduction of oceanic crust occurs along deep sea trench systems.*

spreading centers. Because these plates are interlocked in a complex jigsaw puzzle configuration, *any spreading movement must be accompanied by movement that removes or consumes an equal volume of lithospheric material.* This removal process, called **subduction**, occurs at the sites of deep oceanic **trenches** known as **convergent or subduction zones**. These zones are created when oceanic crust collides with continental crust or when two oceanic plates collide. Essentially, one crustal plates collide and do not subduct, forming huge, uplifted mountain ranges. The Himalayas are an example of such a continent - continent collision.

Paleomagnetism

You may be aware that the earth has a magnetic field similar to that of a dipole magnet. You may not know, however, that the earth's

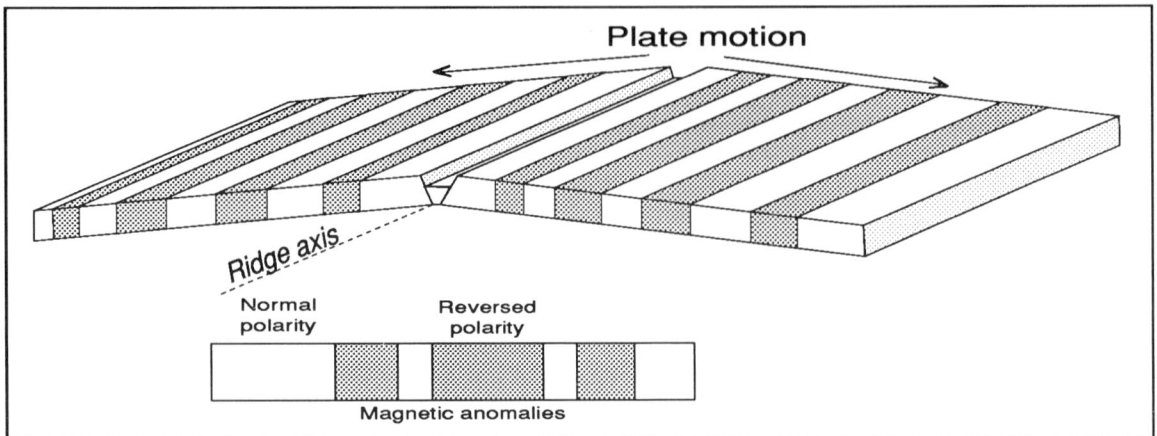

FIGURE 3-2. *Magnetic anomalies. As the sea floor spreads away from the mid-oceanic ridge, new ocean floor is created which records the polarity of the earth's magnetic field. Magnetic anomalies are symmetrical about the ridge.*

magnetic field has reversed its **polarity** (sporadically) many times in the past (approximately every 500,000 years). That is to say, that the present-day north magnetic pole becomes the south magnetic pole.

The polarity of the earth's magnetic field is recorded in igneous rock such as basalt. Igneous rock is formed when hot, liquid rock (magma) cools. As stated earlier, mid-oceanic ridges are divergent zones where two crustal plates are separating. As the plates separate, basalt intrudes into cracks of the ridge forming new oceanic crust. As the basalt cools, the earth's polarity is preserved in magnetic minerals within the rock. Eventually the new crust moves away from the ridge and is replaced by younger basalt which in turn preserves the current polarity (Figure 2). These successive zones of paleomagnetism on the sea floor are known as **magnetic stripes** or **magnetic anomalies** and represent important evidence for sea-floor spreading. If rocks have the same magnetic polarity as the earth's "normal" magnetic field, then a magnetic reading over such a body of rock will have an unusually high value or a positive magnetic anomaly. This is so because the fossil magnetism reinforces the intensity of the earth's present-day normal field. In contrast, rocks that contain a fossil magnetism that is "reversed" in respect to the present-day magnetic field will have a low value or a negative anomaly.

Critical evidence for sea-floor spreading came from the discovery of alternating positive and negative anomalies that were *symmetrical* about the mid-oceanic ridges, coupled with the fact that the ocean floor becomes older as you *move away* from the ridge. Knowing the age of a magnetic anomaly and the distance from the ridge crest, one can determine the *rate* at which the sea floor is being generated. To determine the *spreading rate*, you would need to divide the distance from the axial ridge by the age of the oldest anomaly. This would give you a value for **half-spreading rates**. Since spreading occurs on both sides of the ridge, the **total spreading rate** value would be twice the calculated half-spreading rate value.

Definitions

Asthenosphere. Region of the earth's upper mantle that deforms in a plastic-like manner due to partial melting.

Lava. Liquid rock, or magma, that has reached the surface of the earth.

Lithosphere. Brittle, rigid, outer shell of the earth (approx. 100 km thick) that includes the crust and the uppermost mantle.

Magma. Molten rock within the earth which crystallizes into igneous rock as it cools.

Mid-ocean ridge. Long, continuous mountain ranges found in all oceans. Ocean crust is created by sea-floor spreading and upwelling magma at the crest of each ridge.

Polarity. The property of a physical system which has two points with opposing characteristics, such as opposite charges or electric potentials.

Subduction zone. A collision plate boundary where one lithospheric plate overrides another. A deep-sea trench is formed in the process.

Transform fault. A boundary separating two lithospheric plates where lateral slippage occurs.

Map A

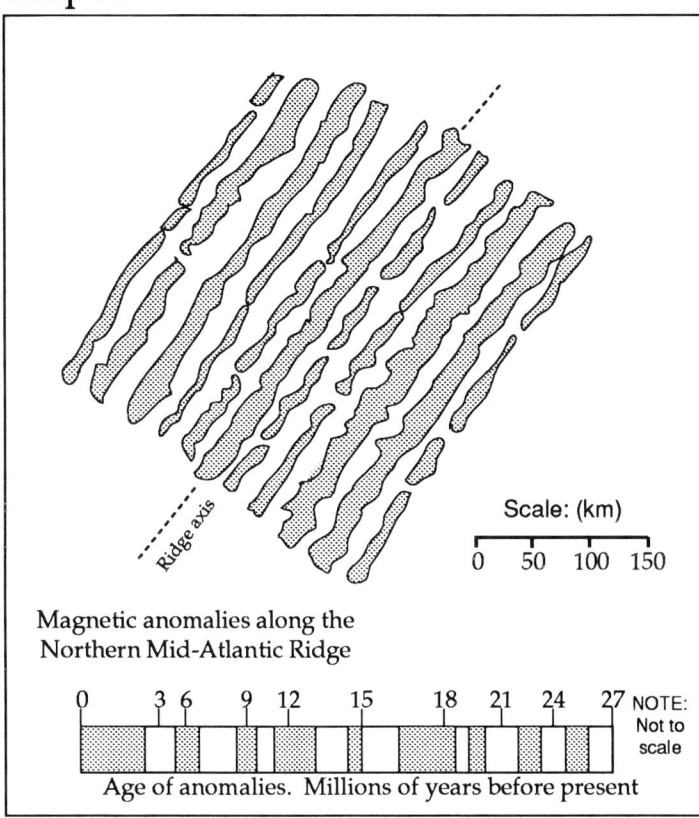

Magnetic anomalies along the Northern Mid-Atlantic Ridge

Age of anomalies. Millions of years before present

NOTE: Not to scale

Exercise

1. Map A shows a series of magnetic anomalies and their corresponding ages that were measured along the *northern* Mid-Atlantic Ridge. Determine the rate at which the northern Atlantic basin has been opening. Do this by making several measurements and taking an average.
   (**Do not use the age scale for measurements!**)

   Express your answer in **centimeters/year.**

Map B

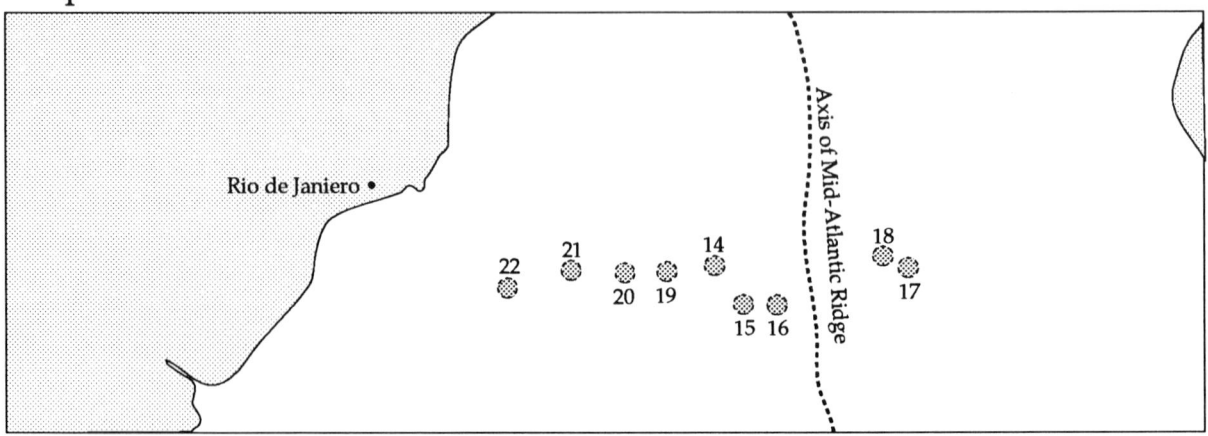

Site	Age of oldest sediment	Distance from ridge
16	11 million years old	280 km
15	24 million years old	400 km
18	26 million years old	490 km
17	33 million years old	630 km
14	40 million years old	720 km
19	49 million years old	1000 km
20	67 million years old	1310 km
21	80 million years old	1680 km

2. Map B shows a series of cores collected across the *South Atlantic* across the Mid-Atlantic Ridge. Determine the average rate at which the southern Atlantic basin has been spreading.
Express your answer in **centimeters/year.**

3. Comparing the two rates, what can you say about the way the Atlantic Ocean basin opened?

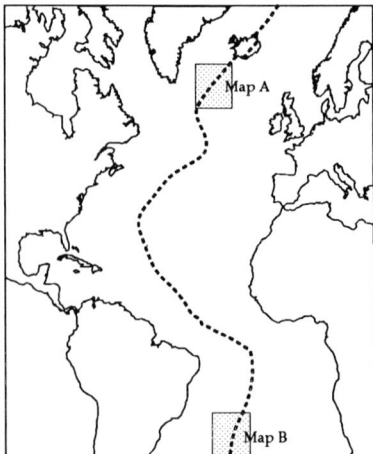

4(a). Plot the age of the oldest sediment against the distance from the the ridge axis for each core. Then, **using a straight edge**, eye-ball the best **straight line** to fit your data. **Calculate the slope of the line.** Slope equals $\Delta Y/\Delta X$ and in this case will represent the average rate of sea-floor spreading for the southern Atlantic basin. Your Y axis is equivalent to kilometers, and your X axis is equivalent to millions of years, thus your answer will have the unit kilometers/million years.

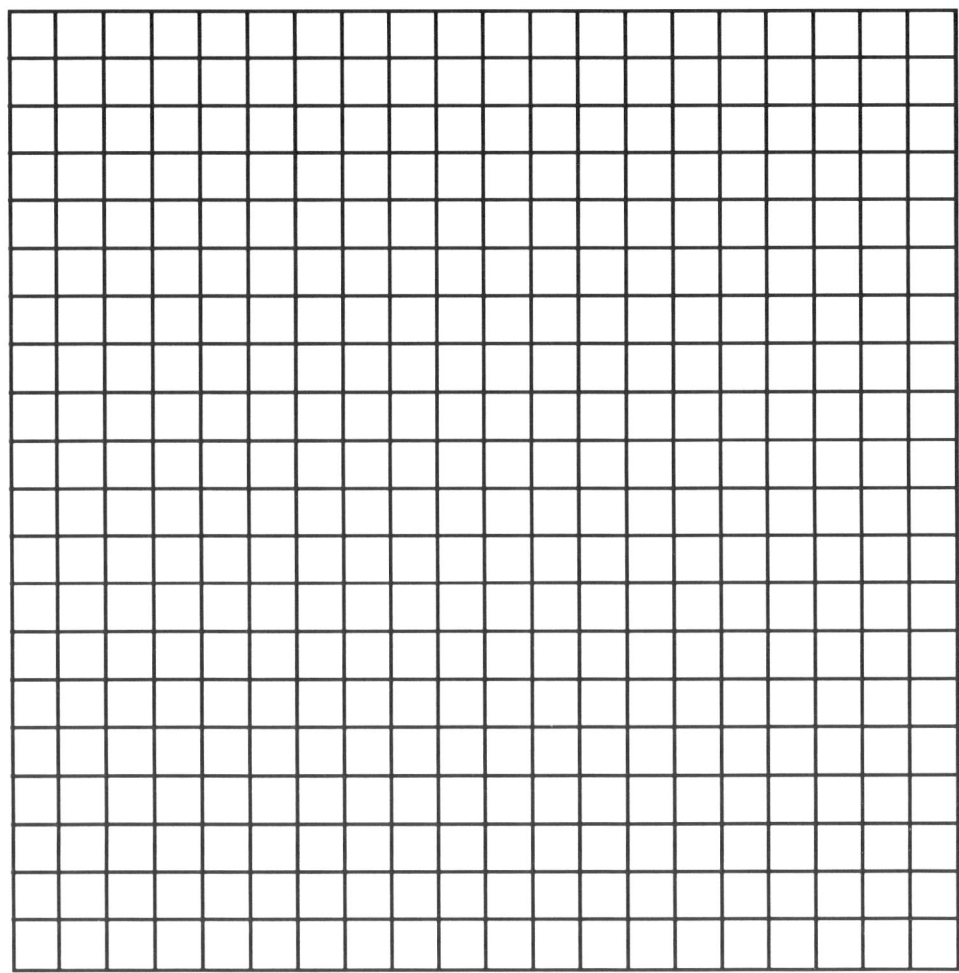

Distance from ridge (km)

Age in million years

(b). Express your answer in terms of **centimeters/year**.

(c). Your answer is a **half-spreading rate**. What is the **total** spreading rate?

5. The distance from the crest of the Mid-Atlantic ridge to the coast of Brazil is about **2700 kilometers**. Calculate from your **half-spreading rate** the approximate length of time which has passed since Brazil was adjacent to Africa.

Map Exercise

Detailed maps of the seafloor will be provided. Carefully analyze the features on the ocean floor and answer the following questions.

Atlantic Ocean

1. Name two places the the Mid-Atlantic Ridge breaks the surface of the sea.

2. What is the greatest depth shown for the Atlantic Ocean? Where is it?

3. Iceland is part of what major feature of the ocean floor?

4. Is the east coast of South America a passive or active margin? How do you know?

5. What makes a better boundary between continental and oceanic areas: the coastline or the shelf break?

6. How does the shape of the Mid-Atlantic Ridge relate to the shape of the continental margins on either side?

7. Note the Kelvin Seamounts (east of Cape Cod). What is their origin? Find other examples of similar origin.

Indian Ocean

1. Is Madagascar part of the African plate? Why or why not?

2. Is Madagascar part of the continent of Africa? Why or why not?

3. List the features which define the boundaries of the Indian plate.

Pacific Ocean

1. The East Pacific Rise is the divergent zone in the Pacific Ocean. What happens to it when it reaches the coast of North America (east of Baja, Mexico)?

2. Where does the ridge reappear?

3. Locate the plate containing Australia. Where is it forming? What is occurring at the leading edge of the Australian Plate?

4. Notice the bend in the Hawaiian-Emperor Seamount Chain. Explain how the bend was formed and why the seamounts become progressively deeper as one moves west and north of Hawaii. **Draw a simple diagram outlining the process.**

5. How can you explain the flat-topped nature of Patton and Parkes Seamounts and Walker Guyot located in the Gulf of Alaska? These seamounts were originally volcanic cones.

6. Compare the earthquake and volcano distribution maps with the Pacific Ocean floor map. What relationship exists between the distribution of earthquakes and the major features of the sea floor?

7. Notice the Liine Islands, Cook Islands and Marshall Islands. Compare their tren to that of the Hawaiian Islands. What is the age of the Line Islands relative to the Hawaiian Islands?

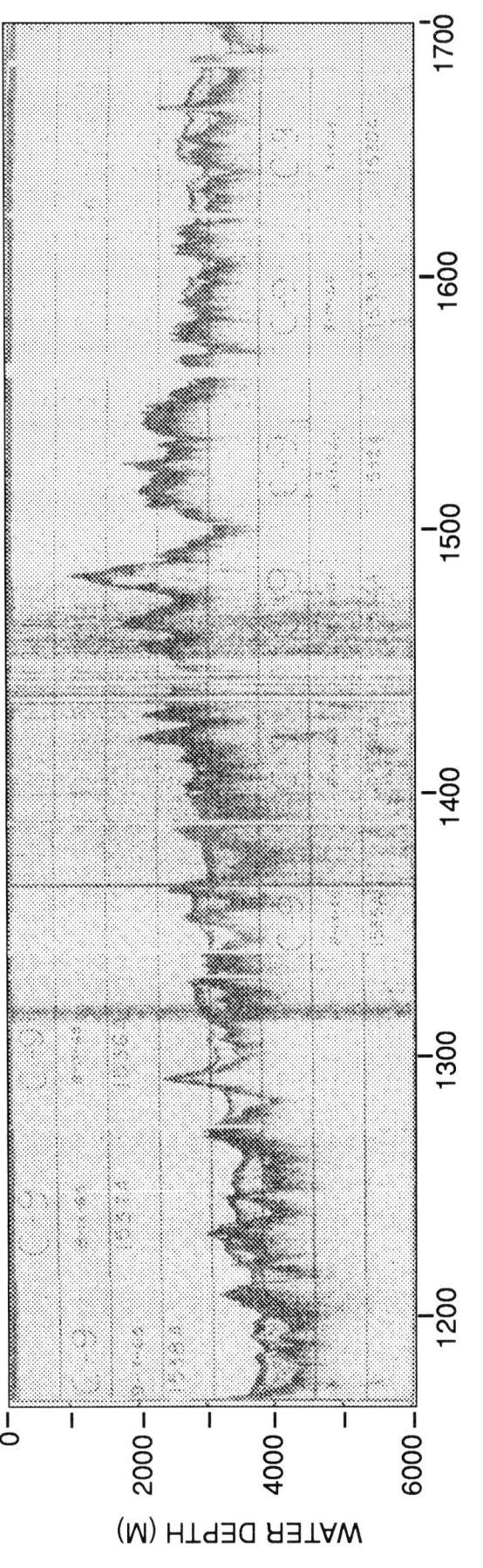

This is a profile of the Mid-Atlantic Ridge.

1. Place an **X** over the axial ridge (The axial ridge should be the area with the greatest local relief).

2. What is the *local relief* of the axial ridge?

3. What is the *total relief* of the profile?

4. What is the material filling in the valleys between 1200-1350 km and 1600 km and on?

5. Why is this material absent between 1350km and 1600 km?

6. What is the vertical exaggeration of the profile? Show all work.

Additional Questions for Review

1. Why are the mid-ocean ridges higher than the surrounding ocean basins?

2. Why are continents higher than the ocean floor?

3. Give 5 arguments supporting continental drift.

4. What stresses (tensile, compression, shearing) characterize the following plate boundaries?

    (a) Transform fault

    (b) Subduction Zone

    (c) Divergent Zone

5. Sketch a transform fault, showing the direction of plate movement.

6. Sketch a cross-section of the lithosphere and asthenosphere, showing areas of accretion and subduction.

7. Why do you think there are so many active volcanoes near subduction zones?

WAVES

Waves have a number of properties which may be calculated in order to learn more about the wave. These properties include:

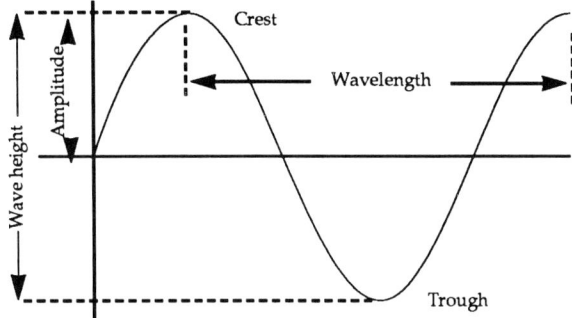

Wavelength (L). The horizontal distance from wave crest to crest or trough to trough.

Wave height (H) The vertical distance between wave crest and trough.

Amplitude (a) Equal to half the wave height.

Period (T) The time it takes for two successive crests or troughs to pass a given point.

Frequency (f) This is the inverse of the period(1/T), and is defined as the number of peaks (or troughs) to pass a fixed point per second.

Wave velocity and motion

The energy or power of a wave determines how vigorously it will attack the coast, and, accordingly, how rapidly it can erode and reshape the shore. Wave energy is a combination of the *velocity* with which the wave strikes the shore, and the *height of the wave*.

The *velocity* or **celerity** (c) of a wave may be represented by the following equation:

$$c = \frac{\text{distance (cm)}}{\text{time (sec)}}$$

Wave velocity is also somewhat dependent upon water depth (d), and this depth determines whether the wave is a *deep-water wave* or a *shallow-water wave*. When water depth is less than L/20, it is a shallow-water wave. When water depth is greater than L/2, it is a deep-water wave. When water depth falls between L/2 and L/20, it is a special case, and calculations become very complicated. Therefore, we will only be considering waves which are very clearly deep- or shallow-water waves.

For a shallow-water wave, velocity may be calculated using the equation:

$$c^2 = gd$$

where g is the acceleration of gravity, equivalent to 9.81 m/s^2. *It is also represented by the slope of a graph of c^2 plotted against time.* Deep-water wave velocity is represented by the equation:

$$c^2 = \frac{gL}{2(\pi)}$$

where π is the mathematical constant approximately equal to 3.14.

Although waves may appear to be progressing through the water, their motion is actually *orbital*. The motion of individual particles of water in a wave is circular, as can be demonstrated by floating a cork on a wave. The cork will simply bob up and down.

Definitions

Amplitude. The distance from level water surface to tip of wave crest or trough; half the wave-height.

Celerity. The speed at which a wave travels.

Deep-water wave. A wave travelling in water with a depth greater than half the wave-length of the wave.

Frequency. The number of wave crests or troughs to pass a fixed point per second.

Height. The vertical distance between wave crest and trough.

Period. The inverse of the frequency, it is the amount of time it takes for two successive wave crests or troughs to pass a given point.

Shallow-water wave. A wave travelling in water with a depth less than 1/20 the wave length.

Wavelength. The horizontal distance between two successive wave crests or troughs.

Exercise

In this experiment we will use simple 'rocker' wave tanks to measure the velocity of waves, and to establish the factor (or factors) which control wave velocities in shallow water.

Your instructor will divide you into groups and assign each group a rocker tank and a stop-watch. Divide responsibility among yourselves for operating the tank, observing the waves, timing them and recording the data.

The actual experimental technique is simple. With the tank filled to the appropriate level (note the depth scale engraved on the side of the tank), the tank operator will tilt the tank until the end with the brass drain rests on the table top. She should hold the tank in this position until the water comes to rest. Then in a single swift motion the tank should be lowered until the support at the opposite end rests on the table.

(1) Start the stop-watch as soon as the tank starts to move downward.

Lowering the end of the tank will start a wave moving back and forth along its length. Allow the crest to make **ten trips** the length of the tank.

(2) At the end of the tenth crossing, stop the watch and record the time required to travel the total distance. Take at least three measurements at each water depth, and calculate the average time. Each student should record the data in the spaces provided on the work sheet.

(3) Calculate the wave velocity for each depth.

Additional information

(i) The inside length of the rocker tank is **120 centimeters.** If you timed 5 crossings of the tank, then the wave traveled a total distance of 600 centimeters.

(ii) **Velocity = Distance/Time**

(iii) **The wave's wavelength is twice the length of the tank.** The wavelength generated in the rocker tank is fixed by the length of the tank - when the tank is tilted the water is high at one end (crest) and low at the other (trough).

Depth (cm)	Travel time for _____ crossings					c (cm/sec)	c² (cm²/sec²)
1 cm							
2 cm							
3 cm							
4 cm							
5 cm							
6 cm							
7 cm							
8 cm							
9 cm							
10 cm							
11 cm							
12 cm							
13 cm							
14 cm							
15 cm							

Plotting

Visual representations of the data in the form of graphs often make it easier to see the results of one's experiments. Plotting some function of the data may help to refine the relationship. Make your plots on the grids provided for you.

1. **Plot your velocity data from the rocker tank. Draw the best-fit straight line or smooth curve.**

2. What is the relationship between wave velocity and water depth?

3. **Plot c^2 vs. depth. Draw the best fit line or curve.**

4. Fromgraph 2, calculate gravity (g). Show all of your work.
   [Hint: c^2/d can be caluated from the **slope** of your graph]

5(a). Would a shorter wavelength affect the velocities you observed? **Explain why?**
   (Consider $c^2 = gd$ and $c^2 = gL/2\pi$)

(b). How deep would the water in the tank have to be for the wave to behave as a deep-water wave?

6. Why do the velocities associated with depths of 15, 14 and 13 centimeters fall off the plotted line?

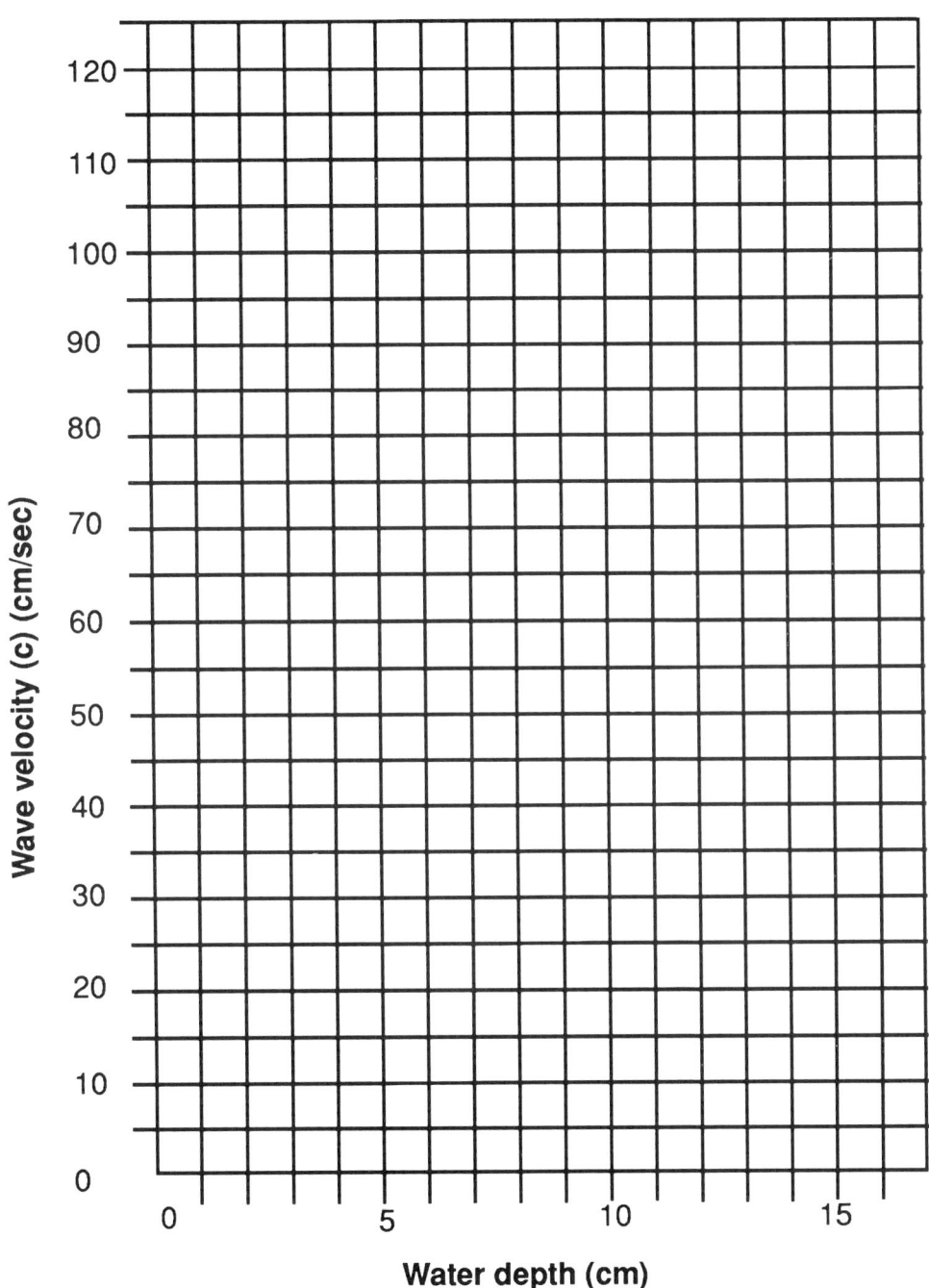

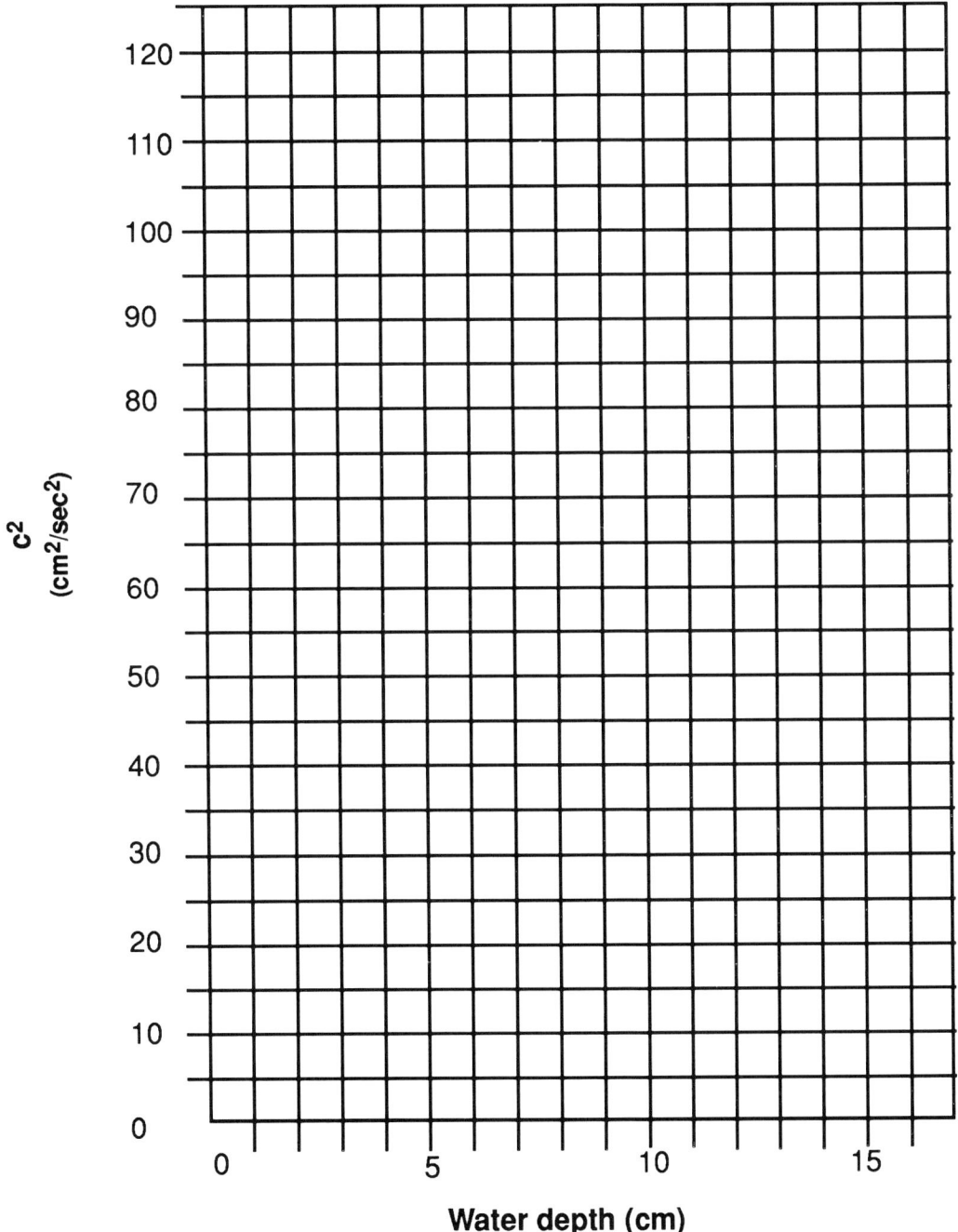

Additional Questions for Review

1. How could you use wave experiments to determine the acceleration of gravity?

2. A wave has a wavelength of 500 cm. How deep would the ocean floor have to be in order for this to be a deep water wave?

3. If a wave has an amplitude of 3 meters, what is the wave height?

4. Define:
   (a) Frequency.

   (b) Period.

THE BEACH ENVIRONMENT

The beach extends landward to the furthest point where sand has been transported by wave action. Above the high water level, *wherever it is horizontal or sloping landward*, is called the **backshore**. The backshore may grade into a dune field or sea cliffs. The *seaward sloping* portion is called the **foreshore**. The foreshore, also called the **swash zone**, *is the region between high and low water levels of the beach*. Primarily, this zone experiences the swash and backwash of breaking waves. A sandy terrace above the high-tide mark, called the **berm**, marks the top of the foreshore as the beach moves into the backshore zone. The berm is formed during high tide. Seaward, beyond the low-tide level, the beach is continuously submerged and is referred to as **offshore**. The offshore may or may not have several troughs and sandbars.

Beaches made up of entirely different components are found in the tropics and on oceanic islands. In these regions a beach sand can be entirely *calcareous* (mainly calcium carbonate; $CaCO_3$) in nature resulting from erosion of coral reefs, marine limestones, and marine invertebrate shells.

Oceanic islands (i.e the Hawaiian islands) may also have beaches composed of volcanic sands. **Olivine** ($FeMgSiO_3$) is a dark green mineral common in basalts and can be found as the major constituent along some beaches. Basalt (the material that forms Hawaiian islands) weathers to a fine beach sand usually black to dark green in color.

A general concept to remember is that the sand supply to the beach in a particular area usually comes from the weathering of geologic formations located inland, sometimes many miles. Rivers and streams bring their sediment load to

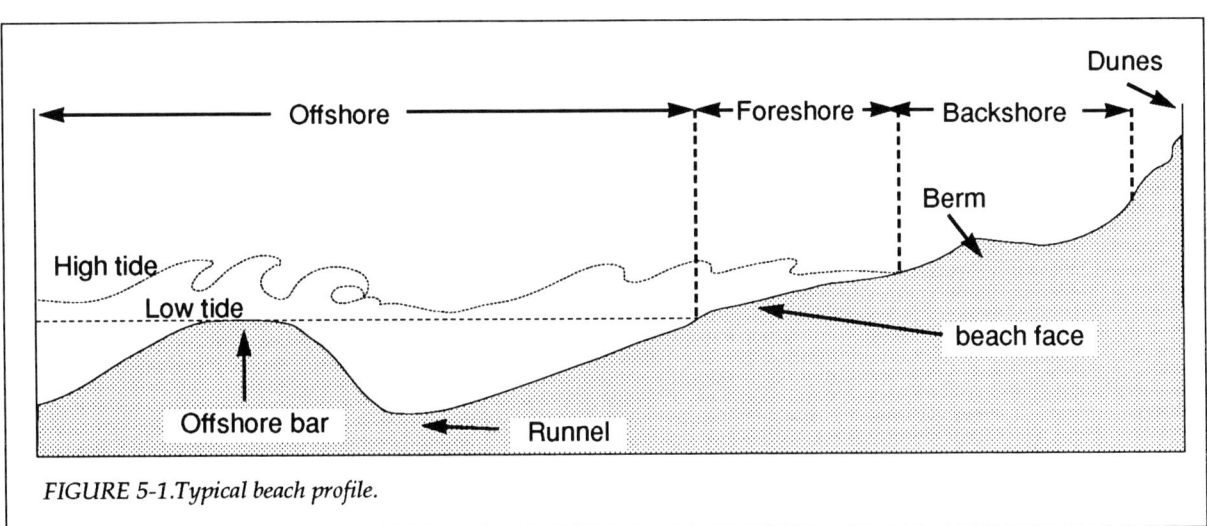

FIGURE 5-1. Typical beach profile.

Composition of beach sands

Beach sands are made up of wide variety of components. **Quartz** (SiO_2) is the most abundant beach sand because it has a high resistance to chemical weathering and wave erosion. Other minerals found in beach sands include **feldspars** and **micas** (K,Na,Mg aluminosilicates), **garnet**, and heavy minerals. **Magnetite** (Fe_3O_4) and **ilmenite** (Titanium oxide) are heavy minerals frequently present. They can be found in enough concentration at some locations that they are 'mined' from the beach. Beaches with these components consist mainly of fine to medium sand. This results in a hard-packed foreshore with a gentle slope.

the ocean where it is distributed to the beaches by wave action and prevailing currents. In addition to this supply source, some sands can be brought in from offshore (such as calcareous sands) by the pounding of waves on coral reefs and seashells.

Methods of sediment determination

By studying the mineralogy of a sediment or sand in detail, one can learn much about the sediment and its source. When the source area or source material of a sediment is being determined, this is known as **provenance**. Usually all minerals in a sand are studied, but sometimes the least common mineral present gives us clues of provenance. Different igneous, sedimentary, and metamorphic rocks have different and distinct accessory minerals.

By using a technique called Heavy Liquid Determination, mineral grains can be separated. Common minerals such as quartz, calcite, and feldspars, have a low specific gravity (= 2.7) and are termed "lights." By putting them in a liquid with a greater specific gravity (bromoform) they will float on the surface. Minerals with a high specific gravity (+2.9) will sink to the bottom in a similar liquid. Therefore, the "light" minerals can be separated from the "heavies." There is some danger here as most heavy liquids for this type of separation are both volatile and toxic.

Minerals with different susceptibility to magnetism can be separated with a Franz Isodynamic Separator. A sample is run through the Franz at a given setting. By running the non-magnetic portion through again at an increased setting and repeating this several times, the sand or sediment can be separated.

Sand size, shape, and sorting

The **shape** of a sand grain can be identified by its *roundness* or *angularity*. A sand grain that has travelled a great distance tends to be more rounded because of abrasion than a grain that has not gone as far. There are of course degrees of roundness or angularity in between these two cases. Another term that is related is sphericity, which is how close the shape of a grain comes to a sphere.

Sand size determines the steepness of the beach face slope. The larger the grain size, the steeper the slopes. Mean size is also a function of distance from the source.

Sorting of sand grains is another important criteria. *Sorting is the measure of how uniform or varied or mixed the sand particle sizes are in any given sample* A beach may have grains that are *well sorted* (all one size in an area) or *poorly sorted* (grains of all sizes intermixed.) We can determine sorting through a *histogram plot*. By sieving an amount of beach sand and then determining the percent of the total amount that is caught in each sieve size, one can plot the percent of the total amount of the sample initially used against each sieve interval. A histogram can tell us the average particle size or a beach.

Composition of sands from various source rocks

Source rock:
Basalt: Forms oceanic crust and hot spot volcanoes on continents and the ocean floor.

Sands:
Basalt grains - common in volcanic settings.
Olivine - dark green mineral common in basalt.
Pyroxene - dark mineral in basalt. Uncommon as a sand. Chemically unstable.

Source rock:
Granite (igneous) and **Schist** (metamorphic): Typical of continental crust.
Sands:
Quartz - common mineral and chemically stable
Feldspar (white or pink) - common, but less stable than quartz.
Amphibole - dark mineral. Chemically unstable.
Mica - Physically breaks down to clay and silt size particles. Tends to wash out to the continental shelf.

Source rock:
Reef material, sea shells
Sands:
Carbonate (calcareous) sand

Definitions

Calcareous. Composed of calcium carbonate.

Backshore. Region above the high tide mark on a beach.

Berm. The flat accumulation of sand on a beach above the high-tide line. Formed by unusually high storm waves.

Foraminifera. Planktonic and benthic protozoans that have a calcareous test (shell).

Foreshore. Region between high and low tide marks on a beach.

Igneous rock. Rock that has crystallized from molten magma.

Metamorphic rock. Igneous, sedimentary, or metamorphic rocks themselves that have been recrystallized by high temperatures and pressures.

Sedimentary rocks. Rocks that have been formed by compaction and cementation of sediments.

Shoal. A submerged bank or bar that is often a navigational hazard. *Shoaling* is the creation of shoals.

Wentworth Grain Size Scale

Grade	Size Diameter (mm)	
Boulders		
	256	
Cobbles		Gravel
	64	
Pebbles		
---------	---- 4 ----------	---------
Granules		
---------	---- 2 ----------	---------
Very coarse		
	1	
Coarse		
	0.5	Sand
Medium		
	0.25	
Fine		
	0.125	
Very fine		
---------	---- 0.0625 ----	---------
Silt		Mud
---------	---- 0.0039 ----	---------
Clay		

Average beach face slopes compared to sediment diameters

Type of sand	Size of particle	Average slope
Very fine	0.0625-0.125	1.0°
Fine	0.125-0.25	3.0°
Medium	0.25-0.5	5.0°
Coarse	0.5-1	7.0°
Very coarse	1 - 2	9.0°
Granules	2 - 4	11.0°
Pebbles	4 - 64	17.0°
Cobbles	64 - 256	24.0°

Sand Descriptions

Calcareous sand:

- (A) - extremely fine sand, reef components, light in color. Composed of reef debris: coral, algal fragments, small foraminifera and sea shells.

- (B) - medium to coarse sand. Composed of reef material, coral, with some dark clays and fine material indicating lagoonal or near lagoonal environment.

- (C) - coarse sand composed of reef material, but with a large percent of dark clays, silt, and lagoonal organics. Present also are iron, sulfides, and dark colored minerals. The dirty color indicates influence from a lagoonal environment.

Quartz sand: typical beach sand of east coast of the United States. Composed of quartz grains, feldspars, garnet, and heavy mineral. This sand has a gritty feel compared to the calcareous sands.

Magnetite sand: this sand comes from a volcanic island. It is one of the most abundant oxide minerals in igneous and metamorphic rocks. This material can make up an entire beach on the island.

Gypsum sand: this is not a beach sand, but instead formed under desert conditions. It is nearly pure white in color and is composed of one mineral; gypsum.

Oolitic sand: Oolites are the product of chemical precipitation of $CaCO_3$ and are easily identified by their distinctive spherical shape.

Olivine sand: composed of olivine grains that have weathered from basalt. Dark green to black in color. This material can make up an entire beach.

East Coast of the United States

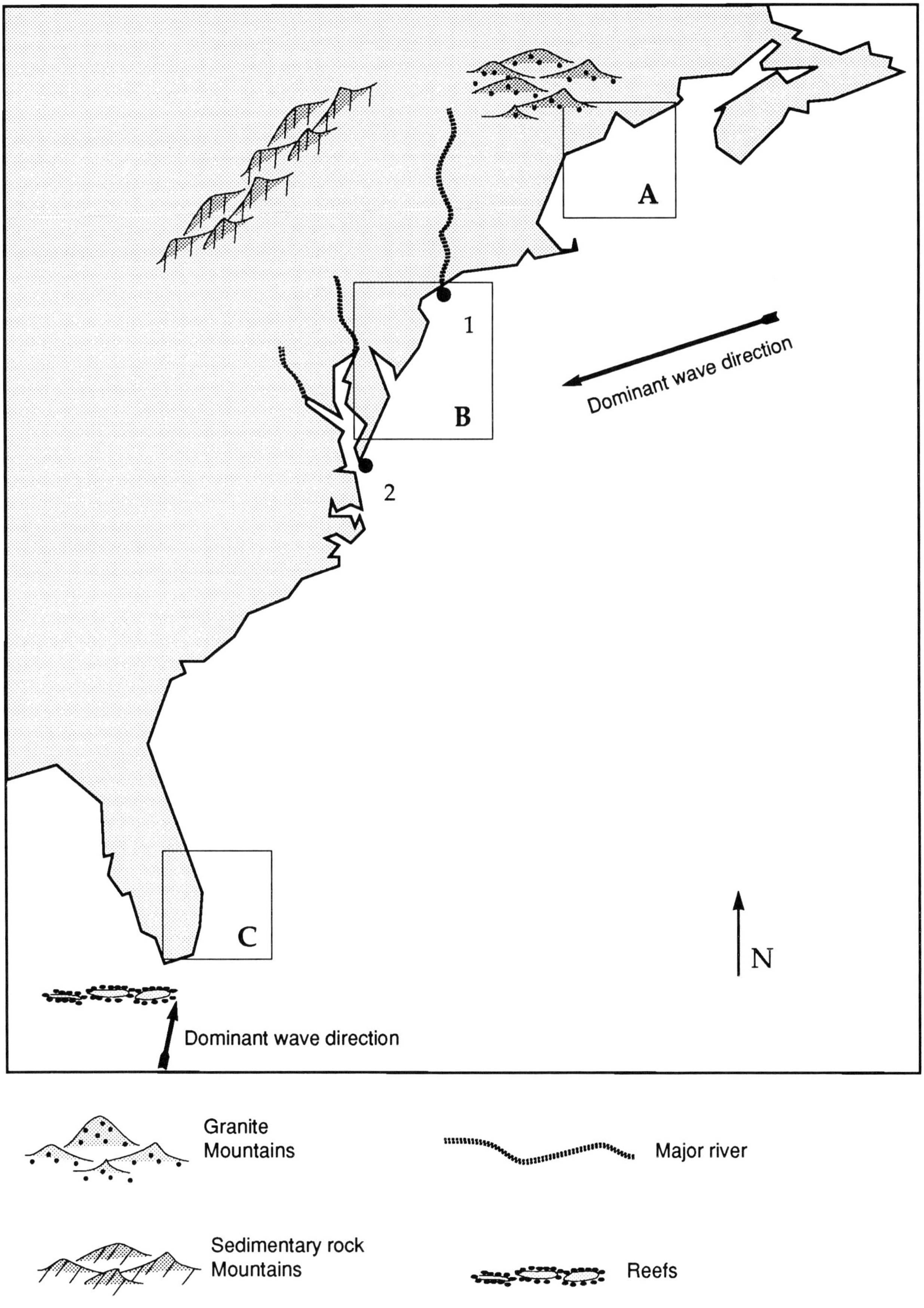

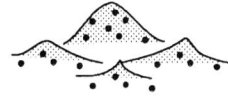 Granite Mountains

 Sedimentary rock Mountains

 Major river

 Reefs

Exercise

Answer the following questions using the map on the previous page.

1. What are the probable compositions for beach sands for location A, B and C on the map.

2. List 5 factors that influence the composition of beach sand at a particular beach.

3. Moving from location 1 to 2, would you expect grain size to increase or decrease? **Explain why.**

4. What factors control grain size and sand distribution at location B?

5. What factors control grain size and sand distribution at location C?

Exercise

Examine the sands given in class under the binocular microscope and answer the following questions.

1. Describe the sands giving:
    a) average grain size (compare to scale)
    b) degree of angularity/roundness
    c) estimate of mineral content
        Quartz (light mineral) vs. heavy minerals (dark minerals)
        Calcareous constituents (does sand fizz?)
        Basalt and related minerals (magnetite)

2. Considering composition, rounding and grain size, rank the samples in order of maturity.

Exercise

Examine the sand samples provided and answer the following questions.

Sample #1. This sand is composed almost entirely of quartz.

1. Describe the rounding.

2. Describe the sorting.

3. Describe its reaction with acid. What does this indicate?

Sample #2. This sand is composed almost entirely of magnetite. Note its magnetic properties.

4. Describe the sorting.

5. Describe its color and luster.

6. Describe the angularity.

Sample #3. This is an oolitic sand.

7. Describe the rounding.

8. Describe its reaction with acid. What does this mean?

Sample #4. This sand is composed of quartz (several colors), feldspar and heavy minerals.

9. Where in the U.S. might this sand be from?

10. What is the source rock?

11. Describe the sorting.

12. Describe its reaction with acid.

Sample #5. This sand is from Hawaii.

13. Name four kinds of particles in the sand.

14. What criteria did you use to identify each particle.

15. Describe the sorting.

Sample #6. This sand is from Hawaii.

16. What is the major component of this sand?

17. Describe the sorting.

18. Describe the rounding.

19. Will the slope of this beach be relatively steep or relatively gentle?

20. Compare **sample #5** with **sample #6**. These samples are both from the island of Hawaii. Explain why there are differences between the two samples.

Sample #7. This sand comes from a volcanic island.

21. Name two kinds of particles in the sand.

22. Will the slope of this beach be relatively steep or relatively gentle?

Sample #8.

23. What is the major component of this sand?

24. Is this sand more likely to be found in a tropical or temperate zone? Why?

25. Describe the sorting.

Additional Questions for Review

1. If you encounter a beach sand with a high content of olivine, what type of source rock would you expect to find?

2. How would you determine if a sand contains particles of calcium carbonate? Magnetite?

3. Describe the appearance of basalt.

4. What type of source rock would you expect to find along an active margin?

BARRIER ISLANDS

If coastal sand supplies are abundant and continental shelf slopes are gentle, the mainland shore can be bordered by a detached system of islands. These islands, known as **barrier islands**, are long and narrow and separated from the mainland by open water. This open water (which could be a lagoon, bay, estuary or salt marsh) can mix with the water of the open shelf by tidal flow, which is directed through a network of **tidal inlets** and **tidal channels**.

The morphology of barrier islands

Barrier islands are distinctively zoned. As one crosses the barrier from its ocean side, the progression of it's environments is:

features. These features are instrumental in creating the back-island flat.

Back-island flat: may be vegetated with terrestrial or marine grasses. This region is above mean high tide.

Salt marsh: a wetland in the intertidal zone. Usually a highly vegetated area (*Spartina* grasses). A salt marsh can be subdivided into low and high tide regions, known as high and low marsh.

Bay-lagoon: a shallow body of water that does not receive a significant amount of freshwater inflow.

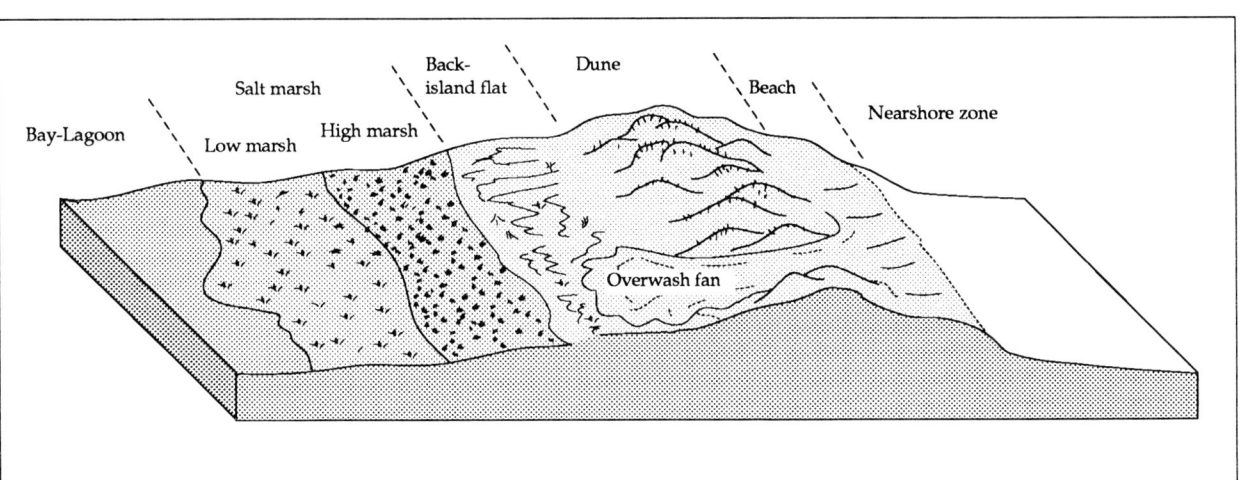

FIGURE 6-1. Barrier island morphology.

Nearshore zone: the zone between the mean high tide and the breaker zone. This a high-energy region that is subjected to longshore (*longshore current* is a current that runs *parallel* to the shoreline) and offshore-onshore transport of sand.

Beach: composed of unconsolidated sand. Above high tide is the *berm* which grades landward into a dune field.

Dune field: dunes may be arranged irregularly or parallel to the shoreline. If they are not stabilized by vegetation, they may move downwind. During periods of intense storms, waves may breach the dune field creating what is known as an *washover fan*.. During these events, sand is carried past the dune field and deposited as lobe-like

Barrier island features

washover fan: a fan-shaped accumulation of sand on the landward side of the barrier island. Created by storm waves.

tidal inlet: A channel that admits the flow of water.

flood tidal delta: a body of sand deposited by the flood tide on the landward side of the barrier island.

ebb tidal delta: a body of sand deposited by the ebb tide on the ocean side of the barrier island.

spit: a narrow tongue of sand extending from the shoreline that is usually created by the longshore drift of sand.

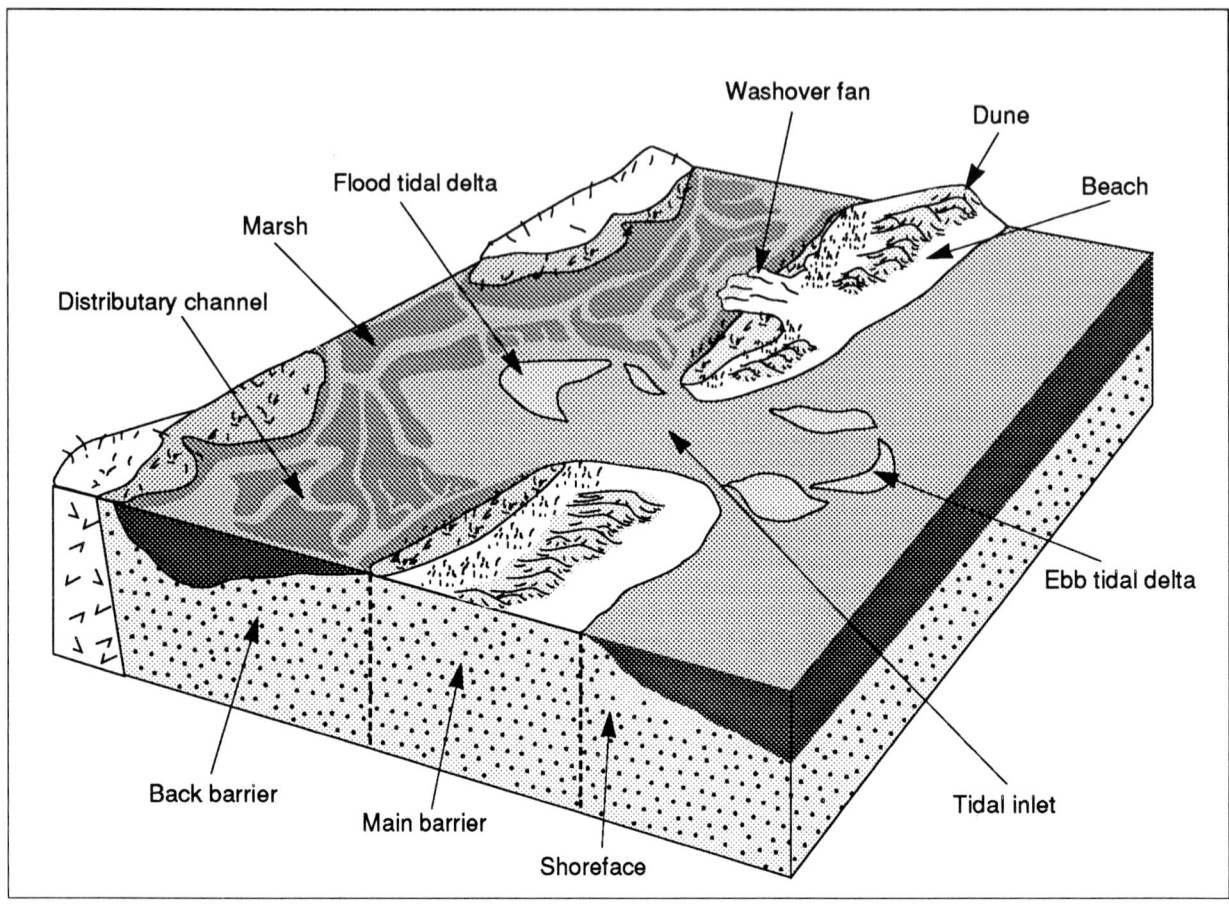

FIGURE 6-2. Barrier island.

Definitions

Berm. The flat accumulation of sand on a beach above the high-tide line.

Offshore bar. A submarine sand ridge in the nearshore zone parallel or subparallel to the shoreline.

Storm surge. An unusually high stand of sea level produced when strong storm winds blow water shoreward.

The Delmarva Peninsula

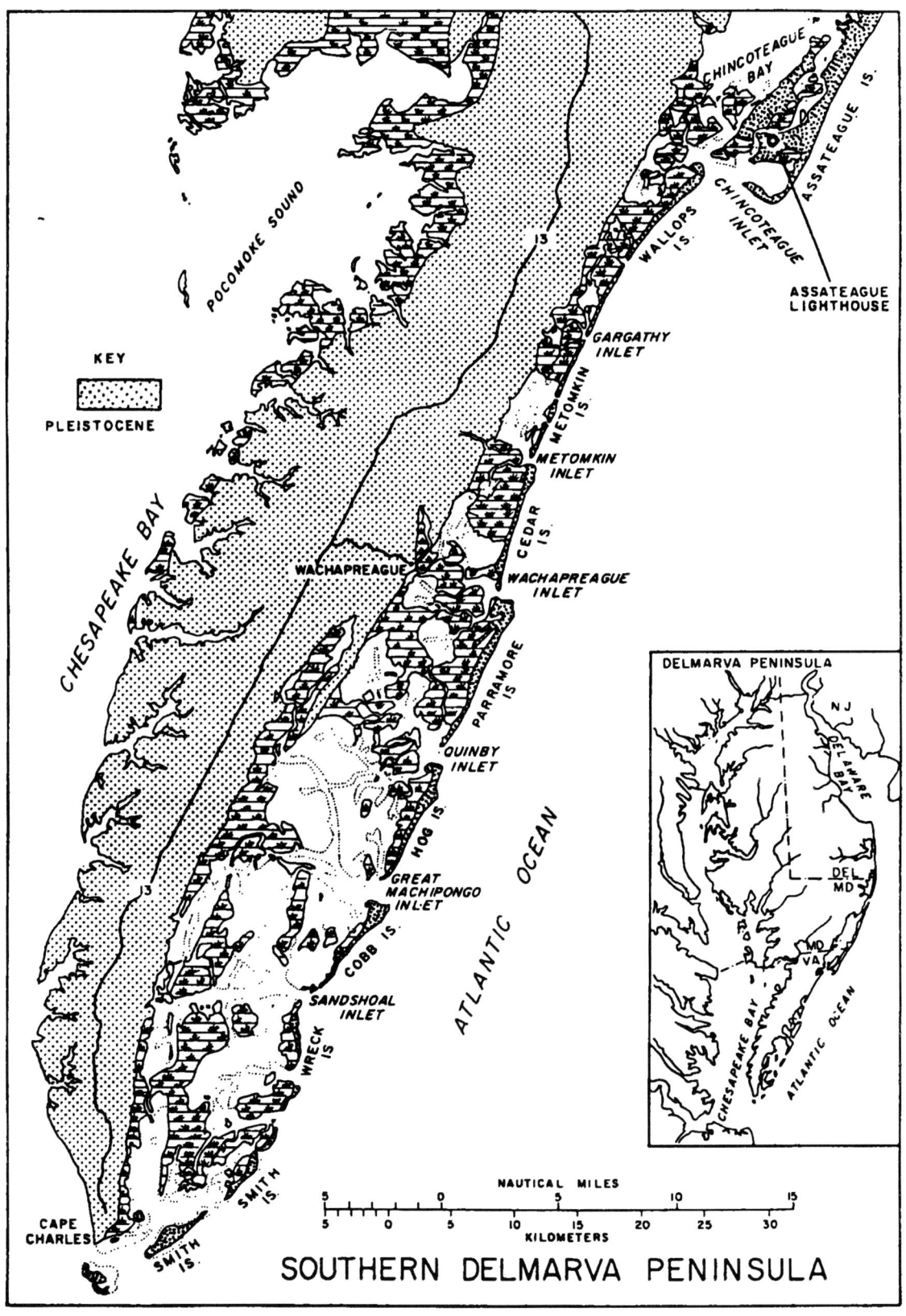

Exercise

Refer to the photographs on pages 125 through 135

The following images are **false color infra-red** photographs of the southern Delmarva peninsula (map is on the previous page) which were taken from an altitude of 10,000 feet.

> **Blue:** Water. (Lighter blue is shallower water with a sandy bottom)
> **White:** Sand or breaking waves
> **Maroon and red:** Vegetation
>   **(1) Maroon-gray:** low marsh
>   **(2) Pink:** High marsh between high and spring high tides. It may also be vegetation on the mainland.
>   **(3) Bright red:** Trees requiring fresh water

1. **Photo #1. Assateague Island**
   What is the origin of the channels which extend from the bay into the back part of the barrier island? (ie: what might they once have been?)

2. **Photo #2. Assateague Island**
   What is the large, funnel-shaped, lavender colored area on this barrier and how did it form?

3. **Photo #3. Assateague Island**
   Are offshore bars found along this stretch of beach?

4. **Photo #3. Assateague Island**
   What is the origin of the large lavender colored feature covering much of the barrier?

5. **Photo #3. Assateague Island**
   Notice the linear features on this barrier. Do you think they are natural or man-made? Considering the recent history of this island (question # 3), what might these linear features be?

6. **Photo #3. Assateague Island**
   Do you see anything in the bay that suggests sediment or sand has been deposited there?

7. **Photo #3. Assateague Island**
   What are the white spots on the photograph?

8. **Photo #4. Assateague Island**
   Do you see evidence of off-shore bars in front of this barrier?
   What is the nature of the evidence?

9. **Photo #4. Assateague Island**
   What is the possible origin of the channel features which run from the bay to the back part of the barrier?

10. **Photo #5. Southern tip of Assateague Island**
    What name is given to such curved features as can be seen on this barrier?

11. **Photo #5. Southern tip of Assateague Island**
    What is the direction of longshore transport?

12. **Photo #5. Southern tip of Assateague Island**
    How do the vegetated ridges relate to the development of this feature?

13. **Photo #6. Southern tip of Wallops Island**
    Notice the inlet. Is there a flood tidal delta? An ebb tidal delta? Which is larger?

14. **Photo #6. Southern tip of Wallops Island**
    Notice the groins along the beach. Which way is the prevailing longshore current moving? Make a sketch of the process.

15. **Photo #6. Southern tip of Wallops Island**
    Notice the color variation in the marsh vegetation. How could you distinguish between the area covered by low tides and those flooded by high tides?

16. **Photo #7. Cedar Island**
   Cite two lines of evidence that indicate this island is migrating landward.

a)

b)

17. **Photo #7. Cedar Island**
   Make a sketch of this photo and **label** as many features as you can.

18. **Photo #8. Metomkin Island**
   What are the dark patches just seaward of the inlets? (Look for the same color elsewhere in the photo).

19. **Photo #8. Metomkin Island**
   From your answer in question #16, what do you infer is happening to these barrier islands?

20. **Photo #9. Quinby Inlet**
   What name is given to the projection of the barrier island in the center of the photograph?

21. **Photo #9. Quinby Inlet**
   Note that sand is accumulating on both sides of this inlet. Sketch the current directions around the inlet which cause this. Include the longshore current and tidal flow on your sketch.
   Label the sketch.

22. **Photo #10. Assawoman Island**
   Cite two lines of evidence to indicate that this barrier island is migrating landward.
a)

b).

23. **Photo #11. Smith Island**
   How do the bright red (vegetated) ridges relate to the development of this section of the island?

SALINITY

Salinity is the *measurement of the total amount of dissolved solids in water*, expressed as parts per million (ppm) (approximately mg/kg) or as parts per thousand (ppt) (approx. g/kg). Typical seawater contains about 35 grams of dissolved solids per kilogram of water, thus the typical marine salinity is about 35 ppt. These dissolved solids are principally:

Chloride (Cl^-) -	55%
Sodium (Na^+) -	30%
Sulfate (SO_4^{2-}) -	8%
Magnesium (Mg^{2+}) -	4%
Calcium (Ca^{2+}) -	1%
Potassium (K^+) -	1%

Additionally, atmospheric gasses, mainly oxygen (O_2) and carbon dioxide (CO_2) are dissolved in

Titration: Titration is a chemical method that measures the amount of chloride within a sample of seawater. The *principle of constant proportion* allows one to determine salinity by measuring one ion in a sample of seawater. The principle states that though the salinity of seawater may vary, the *relative proportion* of the major dissolved ions does not. In other words, the ratio of any two major constituents dissolved in seawater, such as Cl^-/SO_4^{2-} or Na^+/K^+, is a fixed value whether the salinity is 31 ppt or 35 ppt.

Because of constant proportions and because the chloride content is easy to determine, the chloride is measured and salinity determined by the relationship:

Salinity (‰) = 1.8066 x chlorinity (‰)

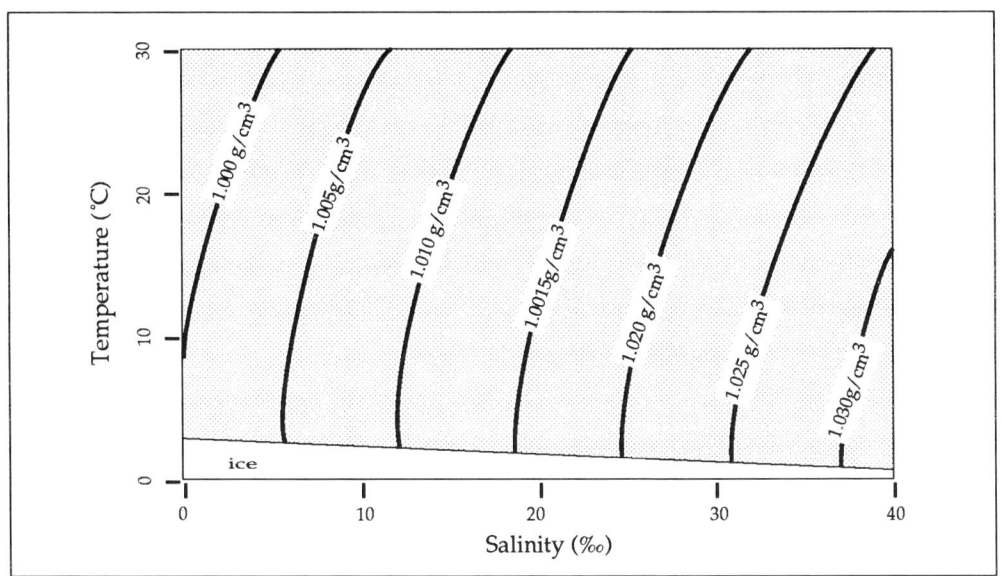

FIGURE 7-1. Plot showing variation of seawater density as a function of temperature and salinity.

sea water. The amount in solution is greatly dependent on temperature and pressure.

Water is classified by its salinity level:

Fresh: 0 - 1 ‰ (‰ is a symbol representing ppt).
Brackish: greater than 1‰, less than 32‰.
Normal marine: between 32‰ and 37‰.
Hypersaline: greater than 37‰.

Measurement of salinity

Measurement of salinity can be made in several ways:

Electrical conductivity of the water: The ability of a liquid to conduct electricity is due, in part, to the amount of dissolved *ions* or *electrolytes*. The salinity of seawater can be determined by it's electrical conductivity using an instrument called a *salinometer.*.

Density: Density is the measurement of the mass within a given volume. As the density (or the number of dissolved particles) of seawater, increases, so does it's salinity. This measurement is made with a *hydrometer*. Density is also controlled by temperature (cooler water is denser than warmer water), which needs to be measured prior to using the hydrometer.

Refraction: Refraction is the angle that light waves (or any waves) are bent as they enter another medium with a different density (or salinity). A *refractometer* converts the degree of refraction (the medium's *refractive index*) into a measurement of salinity.

Contour lines

As you may know, contour lines can be used to show variations of altitude on the surface of the earth. However, they can also be used to show variations in salinity, temperature, pressure, density, etc. Salinity contours may be plotted so that a clear picture of the distribution of high and low salinity waters can be obtained. **Isohaline** lines are contour lines that connect points of equal salinity.

Definitions

Density. The mass of a given substance per unit volume. The closeness of particles or texture in a given material.

Ions. An isolated electron or positron or an atom or molecule which by loss or gain of electrons has acquired a net electric charge.

Isohaline. A line on a chart connecting all points of equal salinity.

Refraction. The change of direction of any wave when it passes through one medium to another in which the wave velocity is different.

Salinity. The measurement of the total amount of dissolved solids in water.

Exercise

Measurement by salinometer

1. As salinity increases so does electrical conductivity (the ease with which an electric current will pass through the water, i.e. the inverse of resistance). This property can be measured with a **salinometer**. You will be given 14 samples of water with different salinities to measure. These samples have been collected at various points in the Chesapeake Bay estuary *from both the bottom water and the surface water..*
   An **estuary** is a semi-enclosed coastal body of water that has both fresh water and marine inputs.

2. Based on your measurements, you will construct salinity contours to obtain an idea of the typical salinity distribution in an estuary.

Chesapeake Bay Salinity Samples	Surface water	Bottom water
Sample 1		
Sample 2		
Sample 3		
Sample 4		
Sample 5		
Sample 6		
Sample 7		

3. Using the measured salinities, draw contours **with 5 ppt contour intervals** for both surface and bottom water. Use different colors to differentiate between top and bottom contours.

KEY:
T = Surface sample B = Bottom sample

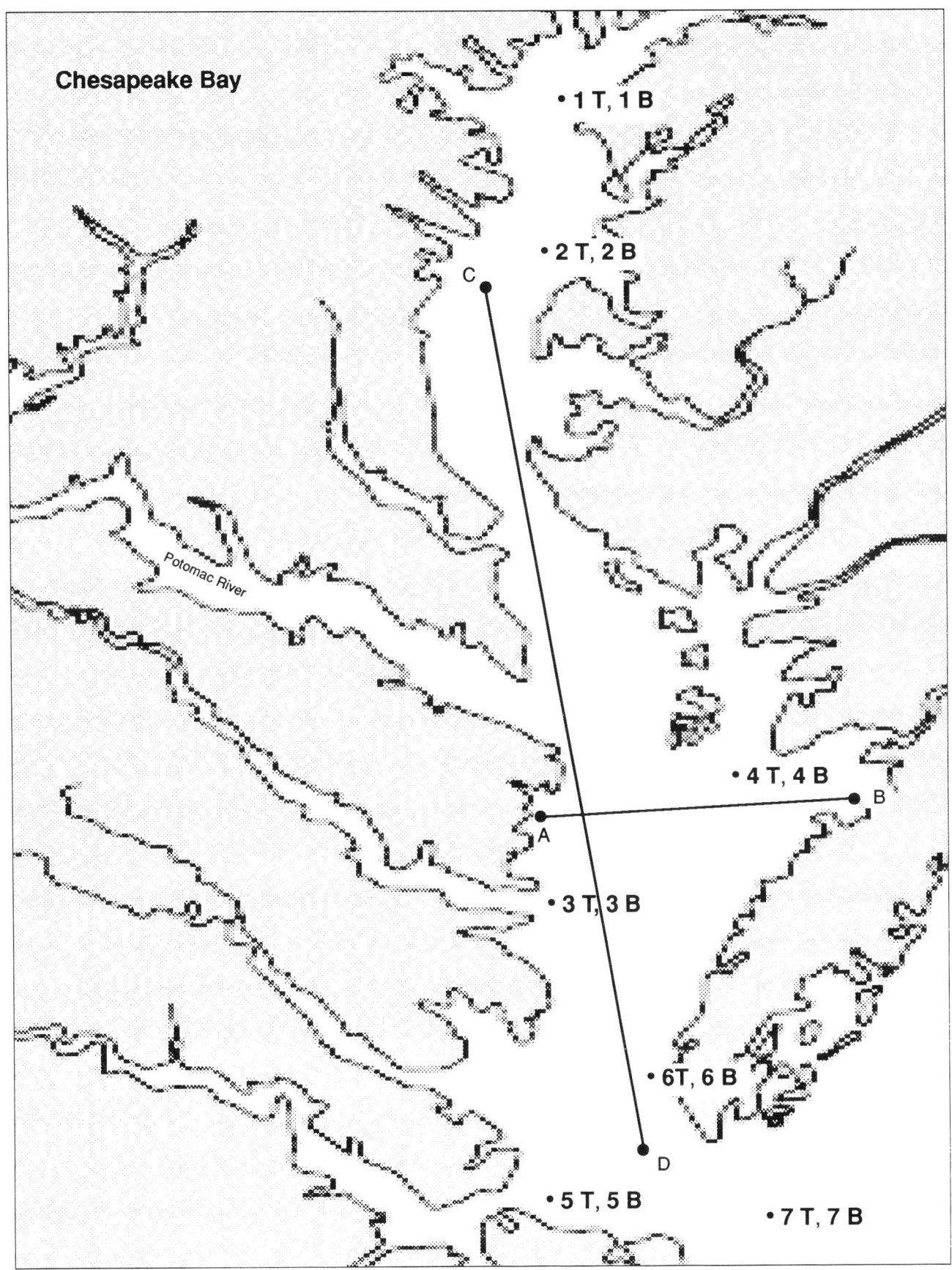

Construct two salinity cross-sections along the lines indicated on the Chesapeake Bay map.

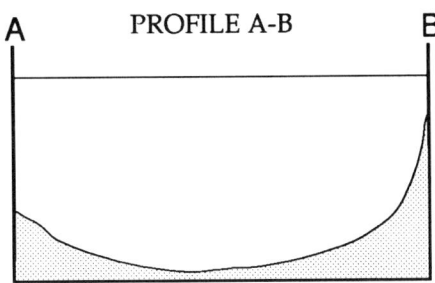

PROFILE A-B

4. **Profile A-B crosses the Chesapeake estuary.**
   To which side of the estuary (east or west) is salinity increasing? **Suggest a reason for this.**
   (Hint: What phenomenon is associated with currents in both the northern and southern hemisphere?)

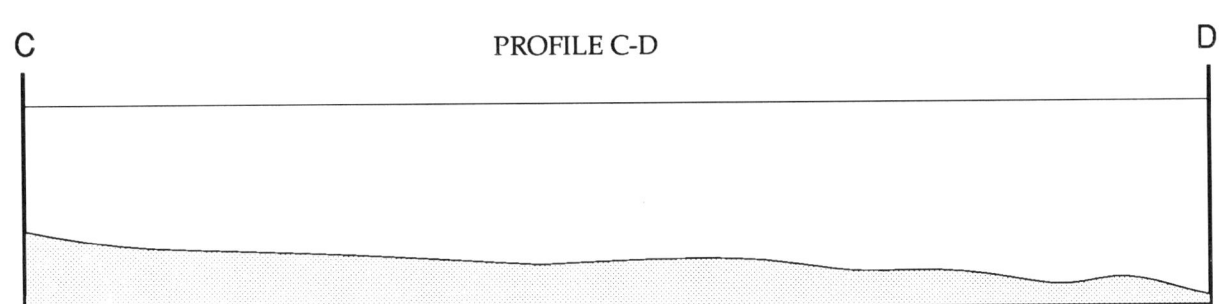

PROFILE C-D

5(a). **Profile C-D.**
   Label the fresh and saltwater bodies on the profile.

(b). What shape and position do they have with respect to each other?

(c). Are these distinct separate bodies or do they merge gradually in to one another?

(d). Explain why these bodies have the shapes, positions and boundaries that they do.

Additional Questions for Review

Draw the salinity contours in the following longitudinal profile of an estuary using contours of **2000 mg/kg (ppm)** intervals.

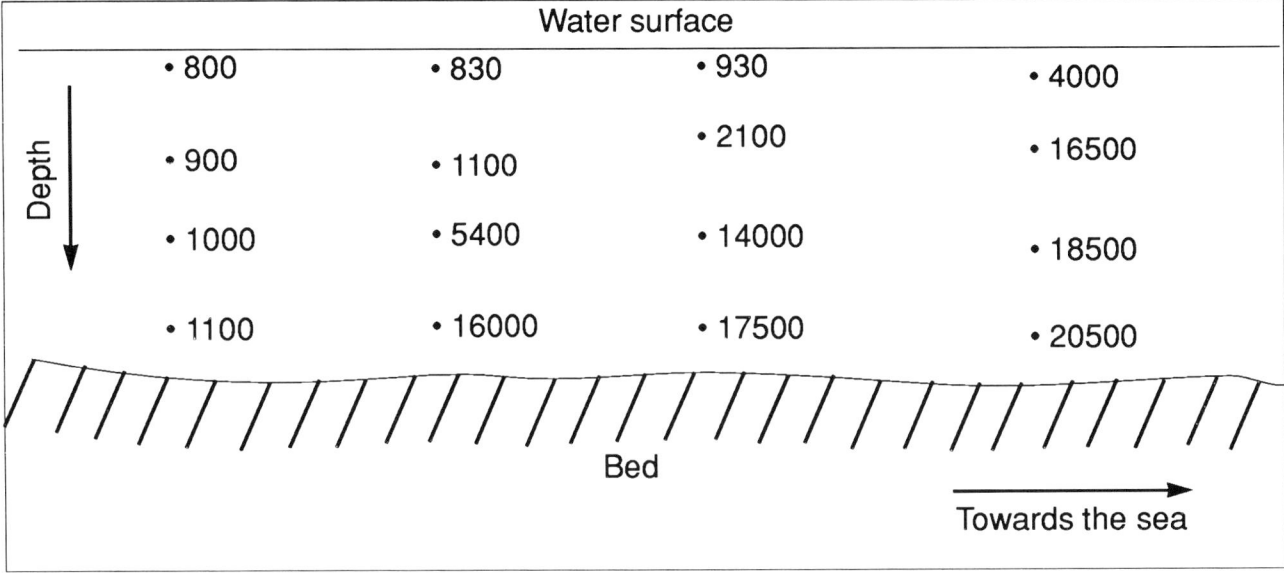

1. How gradually or suddenly does the freshwater merge with the saltwater?

2. Label the fresh and saltwater bodies. What positions and shapes do they have with respect to one another. Why?

3. As well as the flow of freshwater from the river to the sea, the tides cause a twice daily piling-up of the sea into the estuary. What effect will this have on the salinity arrangements above? What implications does this have for organisms such as oysters that inhabit estuaries?

OCEAN STRATIFICATION

Temperature

Three distinct thermal zones or layers characterize the vertical temperature profile of most of the world's oceans. Generally, temperature decreases with depth (fig. 1). However, in low to mid-latitudes a surface layer tens of meters thick (**mixed layer**) is typically **isothermal** or uniform. This is caused by the mixing action or **turbulence** created by winds blowing over the oceans surface. Below this layer, temperature rapidly decreases. This steep temperature gradient, known as the **thermocline**, ranges in water depth between 200 to 1000 m. The thermocline is better developed in summer and more poorly developed in winter due to seasonal temperature variations. Below the thermocline, deep water (**deep layer**) is characteristically isothermal averaging less than 4 °C.

High latitude waters have a temperature profile nearly uniform from surface to deep waters. This is because the cold air temperatures and low light intensity (during winter months) prevent the formation of a thermocline, thus the

tion of evaporation and precipitation. Regions that experience high evaporation rates and low precipitation subsequently have waters with higher salinities than the global average. If precipitation is higher than evaporation then lower salinities can be expected. These processes are controlled by climate, and since climate varies with latitude, so does salinity.

As the thermocline (a steep temperature gradient) represents a boundary between two water masses in the ocean's vertical profile, so does the **halocline** (a steep salinity gradient). As does the thermocline, the halocline tends to **stratify** the ocean into layers. This stratification is most pronounced between 40°N and 40°S latitude with high-salinity surface waters separated by less saline deep waters.

Density

Density is a function of temperature, salinity (see Salinity lab) and pressure. In order for a water column to remain stable, denser waters must be *overlain* by less dense waters.

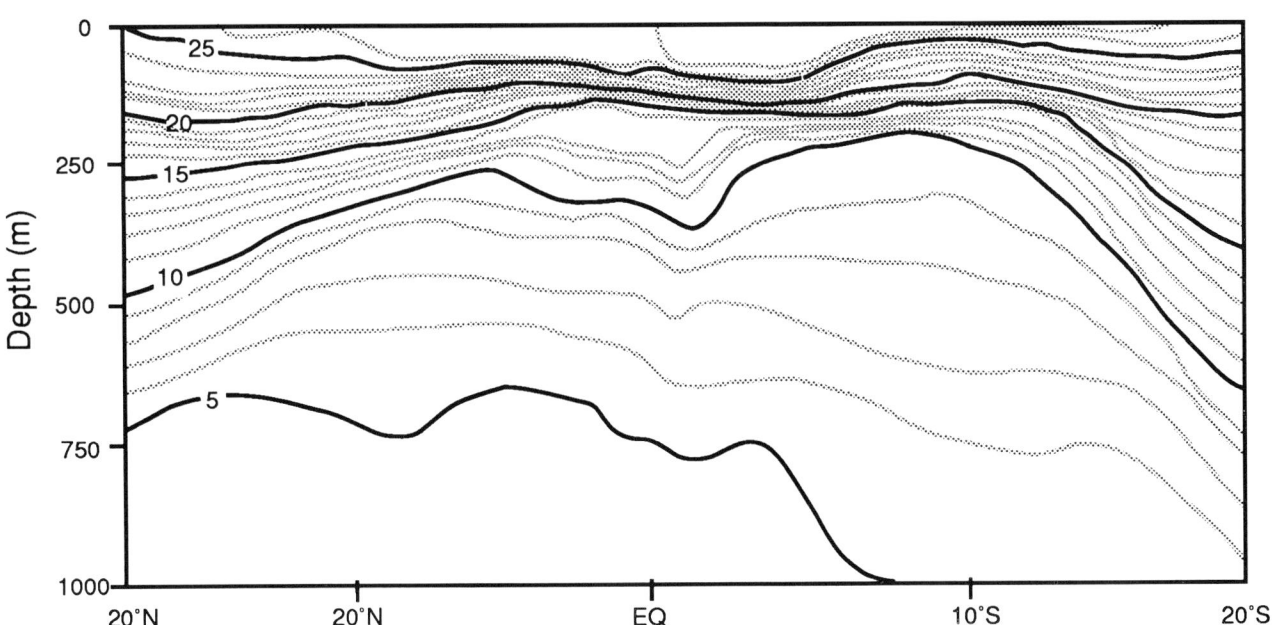

FIGURE 8-1. *Temperature profile of the western Atlantic Ocean. Upper 1000 meters is shown. Average depth is 4500 meters. Isotherms shown in °C.*

waters are well mixed throughout the water column.

Salinity

The average salinity for the oceans is 34.7‰. Variations in salinity are generally a func-

Dense waters can be formed by *decreasing its temperature and increasing its salinity*. However, if low salinity waters are cooled they may become more dense than high-salinity waters with higher temperatures. These low salinity, denser waters would then sink below less dense (possibly higher salinity) water.

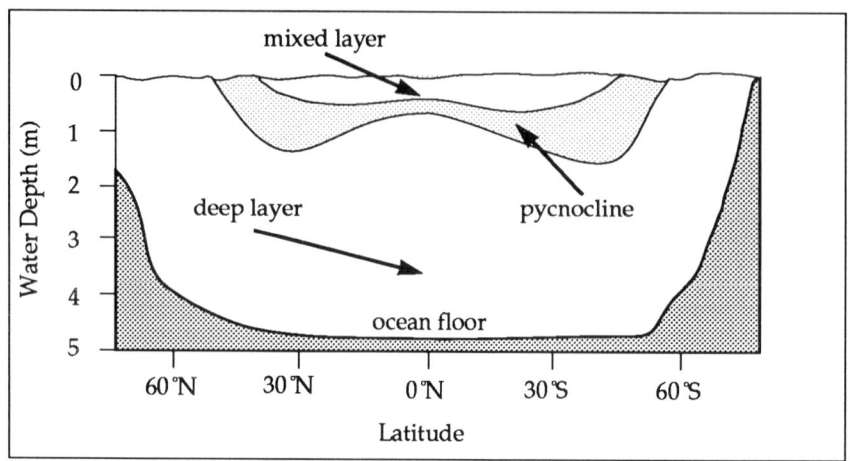

FIGURE 8-2. Density profile of the ocean. Note that pycnocline is not well developed at higher latitudes.

Temperature exerts more control on density than salinity, thus the density profile in the ocean mirrors the temperature profile. Three density zones can be identified: a surface layer, **pycnocline**, and deep layer (fig. 2). Once again, the pycnocline (a steep density gradient) separates two relatively well-mixed water masses. This density stratification is most pronounced in low to mid-latitudes.

Stratification of the water column has important effects on the distribution of nutrients and thus productivity of the oceans. This topic will be explored in more detail in following labs.

Definitions

Isotherm. A line connecting points of equal temperature.

Isothermal. Occurring at constant temperature.

Stratify. To form or become arranged in strata or layers.

Turbulence. Agitated or chaotic movement of water

Exercise
Using the data given below construct three graphs on the following pages.

> **Graph (A) is a temperature profile for a given location in the Atlantic Ocean.**

1. Identify the depth range of the thermocline.

2. Calculate the average change of temperature per unit depth through **each** thermal layer.

3. Which thermal layer (water mass) composes the bulk of oceanic water?

4. What is the origin of the cold deep water?

> **Graph (B) is a salinity profile for the same location in the Atlantic Ocean.**

5. What is the range of salinity variation from the surface to the deep layer?

6. Identify the depth range of the halocline.

> **Graph (C) is a density profile. In order to construct this graph you will need to derive the density of the water at a particular depth by comparing temperature and salinity on the T-S diagram provided.**

7. Identify the depth range of the pycnocline.

8. Explain why there is very weak density stratification at higher latitudes.

The data below were collected from the western Atlantic Ocean.
Site location is 3° 56' N, 38° 31' W.

Depth (m)	Temp (°C)	Salinity (‰)
1	27.637	35.986
36	27.554	35.993
80	27.53	36.032
101	26.943	36.14
126	23.54	36.26
135	21.247	36.304
151	15.21	35.17
199	10.99	35.04
214	10.682	35.006
248	10.03	34.92
296	9.447	34.872
375	8.281	34.754
444	7.537	34.662
519	6.843	34.638
667	5.416	34.548
743	4.995	34.543
811	4.777	34.564
883	4.637	34.594
940	4.540	34.617
999	4.482	34.655
1058	4.524	34.713
1123	4.450	34.733
1174	4.483	34.792
1221	4.485	34.826
1285	4.498	34.885
1301	4.491	34.895
1346	4.394	34.913
1401	4.295	34.392
1552	4.128	34.971
1703	3.877	34.976
1851	3.626	34.969
1992	3.437	34.964
2140	3.210	34.958
2290	3.011	34.949
2741	2.604	34.930
3053	2.395	34.919
3503	2.192	34.913
4056	1.694	34.864

Graph (A)
Temperature Profile (°C)

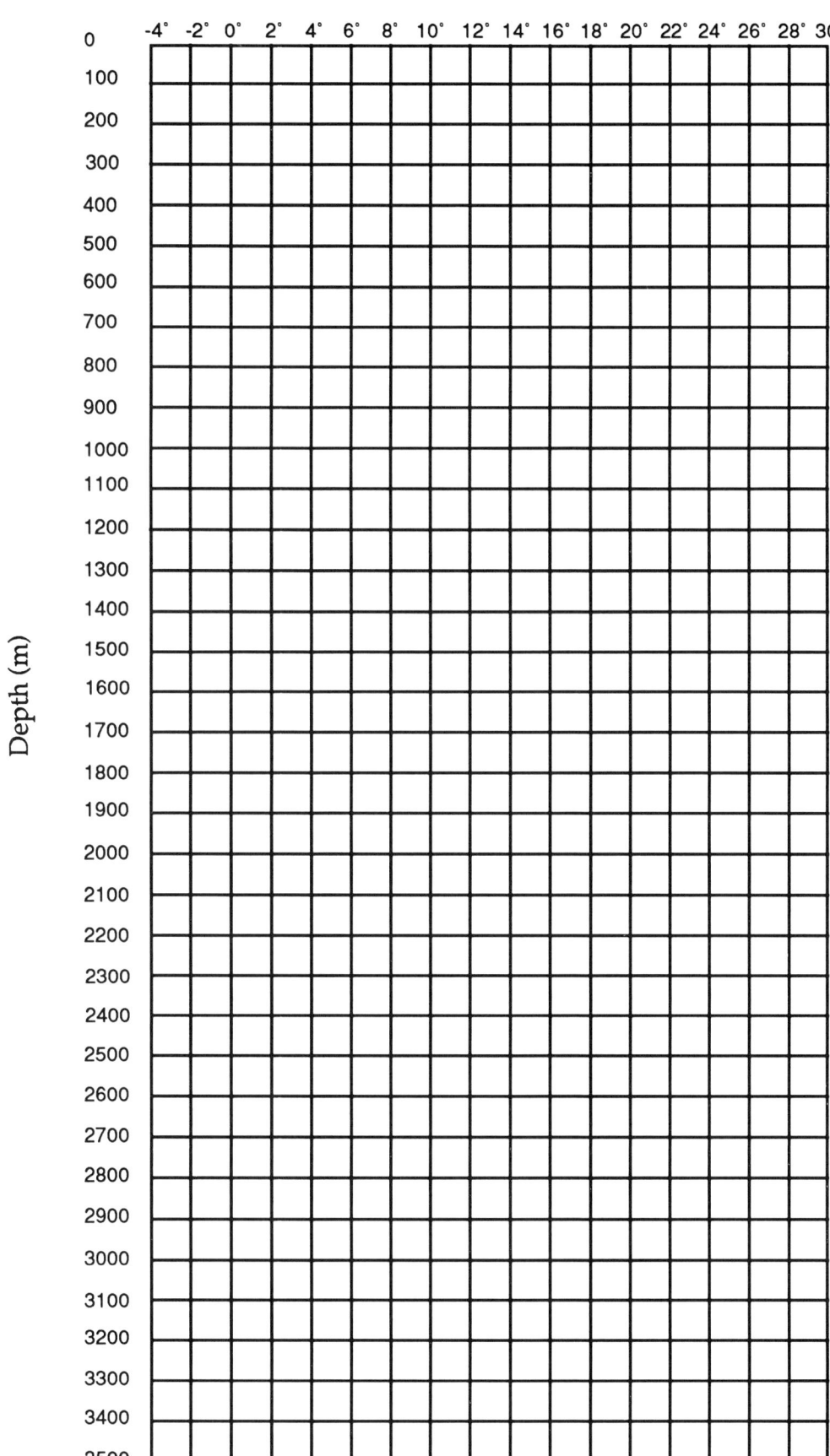

Graph (B)
Salinity Profile (‰)   Salinity (‰)

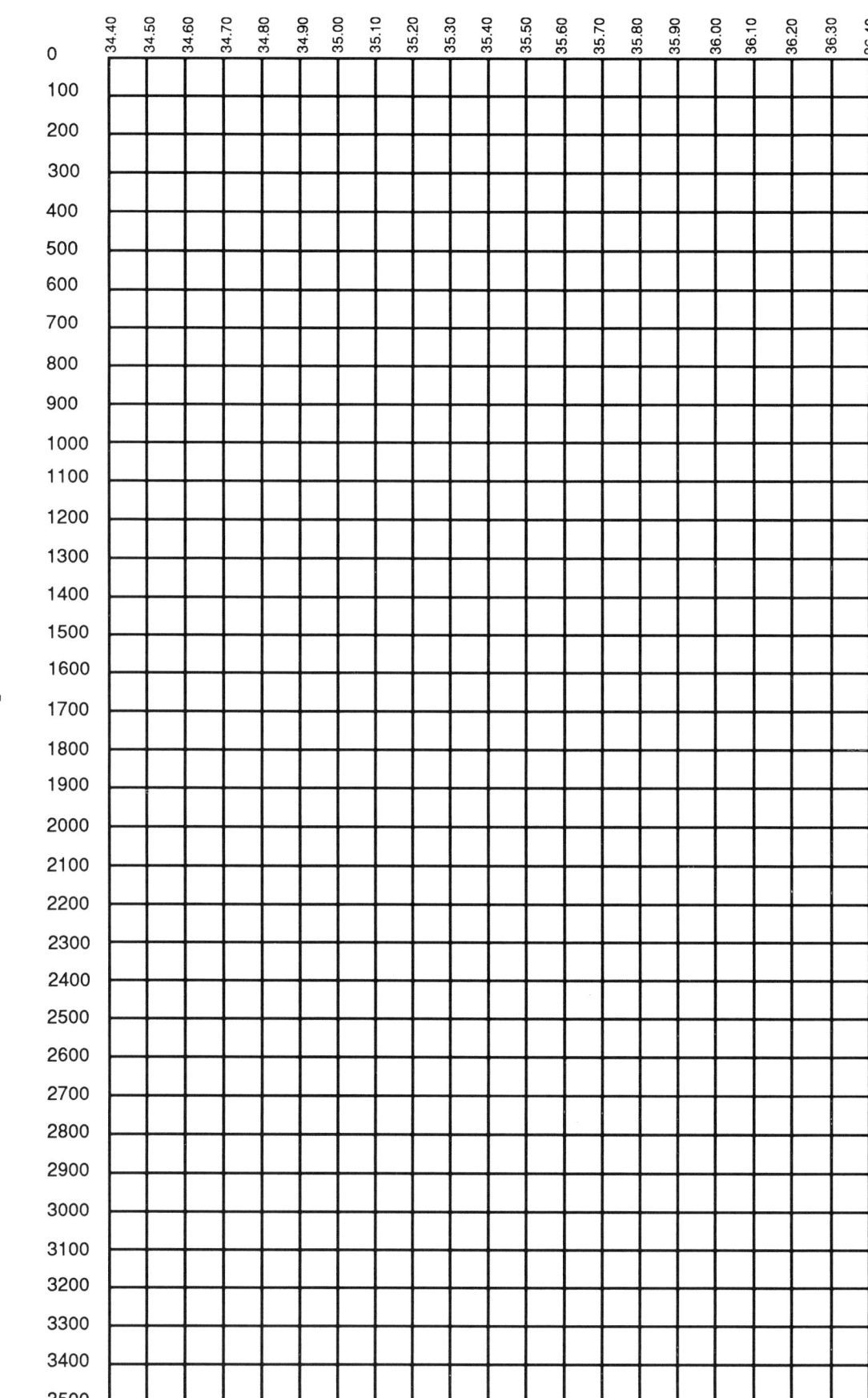

T-S diagram showing sea water density as a function of temperature and salinity

(a water sample whose salinity is 34‰ and whose temperature is 20°C would have a density of 1.024 g/cc.)

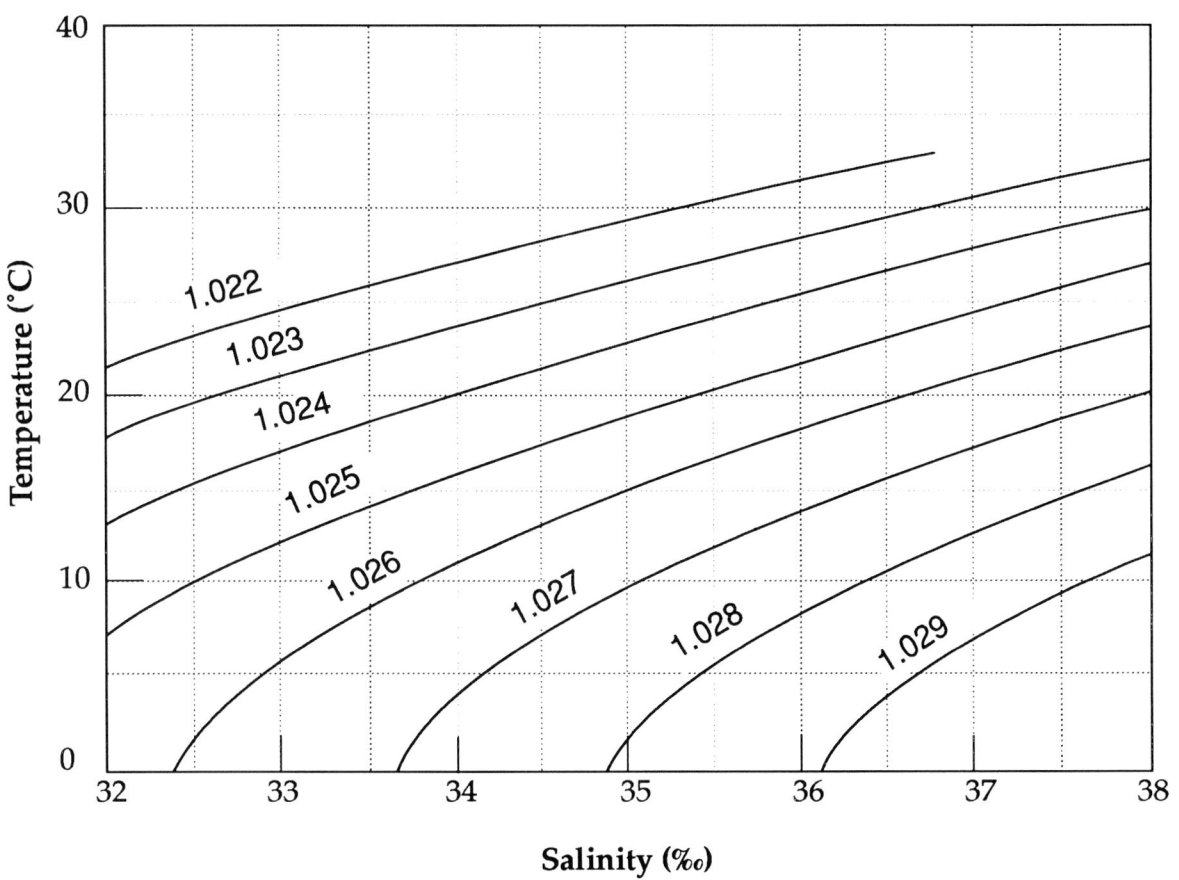

Graph (C)
Density Profile

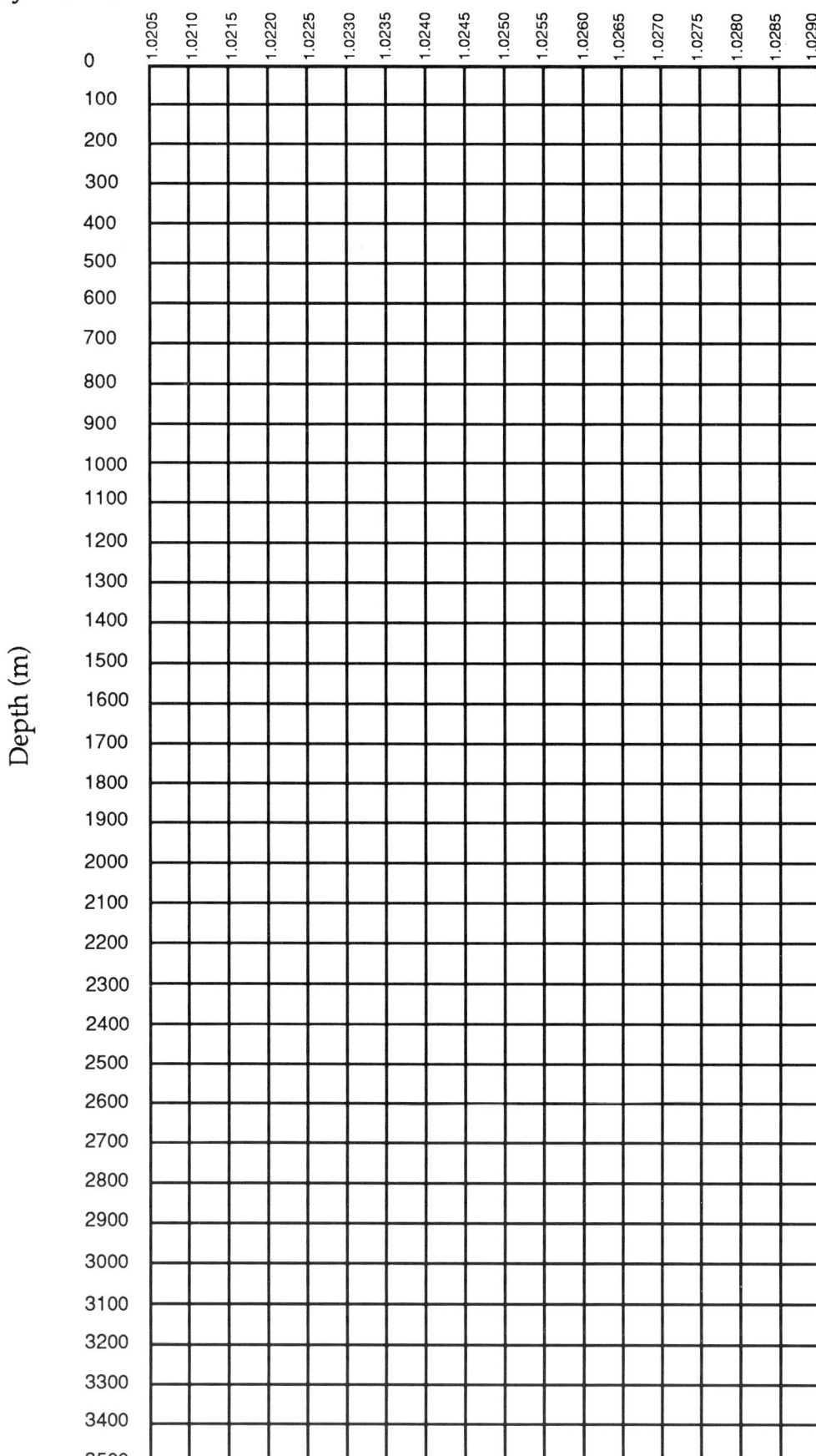

Additional Questions for Review

1. Define thermocline:

    halocline:

    pycnocline:

2. Using figure 8-1:
    (a) Calculate the change of temperature per meter at 20°N from 0 to 250 meters.

    (b) Do the same from 250 to 500 meters.

3. Density stratification slows the vertical mixing of oceanic waters. How would this effect surface water productivity?

OCEAN NUTRIENT DYNAMICS

Carbon, nitrate, phosphorous

Organic matter produced by photosynthesis is the foundation to life on this planet. As you know, photosynthesis is the process by which organisms synthesize food from inorganic elements. This reaction can be represented in very simplified form as:

$$6\ CO_2 + 6\ H_2O + \text{sun energy} \longrightarrow C_6H_{12}O_6 + 6\ CO_2$$

However, carbon dioxide, water, and sunlight are not the only elements that photosynthesizers require in order to produce organic matter. Nutrients such as **nitrogen** (in the form of **nitrate** (NO_3^-), **nitrite** (NO_2^{2-}), and **ammonia** (NH_3)) and **phosphorous** (in the form of **phosphate** (PO_4^{3-})) are also critical to growth. Other nutrients such as **dissolved silica** are vital if the marine plant (such as diatoms) precipitate a silica skeleton. The photosynthetic reaction can be expanded to include these nutrients:

$$106\ CO_2 + 122\ H_2O + 16\ HNO_3 + H_3PO_4 + \text{sun energy}$$
$$\longrightarrow (CH_2O)_{106}\ (NH_3)_{16}\ H_3PO_4 + 138\ O_2$$

nitrate and phosphate, however, are close to zero in surface waters. Because these nutrients are in low supply they tend to limit the extent of phytoplankton **productivity**. Thus they are known as **biolimiting elements**.

Micro-nutrients or **trace metals**, such as iron, manganese, boron, zinc, and copper are also important to productivity.

Oxygen

Phytoplankton productivity occurs mainly within the top 100 meters of the water column. This region where the majority of light is present is known as the **photic zone**. Because plant organic matter is the foundation of the food chain, small to large marine animals also occupy this zone. When these organisms die they sink through the water column and are consumed or slowly decompose. Eventually this organic matter is reduced to its elemental constituents. This process, known as **nutrient regeneration** or **remineralization**, returns nitrogen and phosphorous to their soluble forms (nitrate and phosphate).

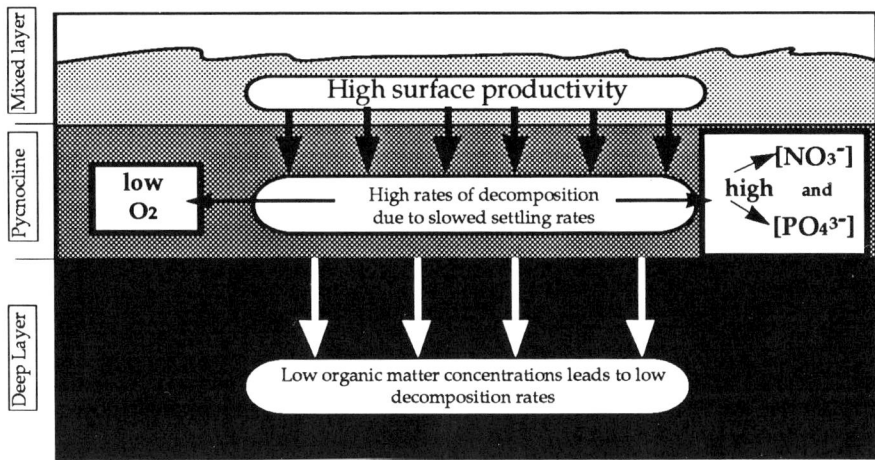

FIGURE 9-1 *Nutrient profile. High rates of decomposition through the pycnocline develops an oxygen minimum zone and high concentrations of dissolved nutrients.*

Marine photosynthesizers (phytoplankton) utilize carbon, nitrogen and phosphorous in predictable proportions. From the above equations it can be seen that these nutrients are extracted from seawater in the ratio of 106 (C): 16 (N): 1 (P). This relationship is known as the **Redfield ratio**.

Carbon (CO_2) in seawater is abundant and readily available to sustain productivity in marine surface waters. The concentration of

Decomposition, however, generally requires oxygen. Thus, as organic matter travels downward through the water column *nitrate and phosphate concentrations increase while dissolved oxygen is depleted.*

In the previous lab you learned that the oceans in low- to mid-latitudes are stratified due to a rapid density gradient known as the pycnocline. This stratification prevents mixing of surface waters with deeper waters. The fall of sinking organic matter is slowed as it passes through the pycnocline, allowing organic material to accumulate in this water layer. High rates of decomposition combined with slow mixing rates in this zone rapidly deplete dissolved oxygen producing an **oxygen minimum layer.**

Definitions

Biolimiting element. An element whose distribution is controlled by biogeochemical processes. These elements are characterized by low concentrations in surface waters and tend to limit phytoplankton growth.

Redfield ratio. The elemental ratio of C to N to P that is present in average phytoplankton (106:16:1).

Remineralization. The degradation of organic matter that leads to the solubilizaton of nutrients.

Soluble. Able to be dissolved in water.

Phytoplankton. Floating, photosynthesizing organisms such as diatoms.

Productivity. Mass of phytoplankton produced per unit time (population growth rate).

Exercise
Using the data given below construct three graphs on the following pages.

> **Graph (A) is a phosphate profile for a given location in the Atlantic Ocean.**

1. What is the phosphate concentration at the surface? Why is it so low?

2. Identify the depth range of maximum phosphate concentration.

> **Graph (B) is a nitrate profile for the same location in the Atlantic Ocean.**

3. What is the nitrate concentration at the surface? Once again, why is it so low?

4. Identify the depth range of maximum nitrate concentration.

5. Compare the distribution of the phosphate curve with that of nitrate. Describe why there are similarities or differences.

6. Calculate the ratio of nitrogen to phosphorous for several depths. What is controlling this ratio?

> **Graph (C) is a oxygen profile.**

7. Identify the depth range of the oxygen minimum zone.

8. Compare the distribution of phosphate, nitrate, and oxygen. Describe the relationship you find and explain why that relationship exists.

9. Compare the density profile from the previous lab with the phosphate, nitrate, and oxygen profiles you have just created. Describe the relationship you see and EXPLAIN why it occurs.

The data below were collected from the western Atlantic Ocean.
Site location is 3° 56' N, 38° 31' W.

	Concentrations in μM/KG		
Depth (m)	Phosphate	Nitrate	Oxygen
1	.07	0.0	189
36	.06	0.0	199
80	.06	0.0	200
135	.041	5.7	163
214	1.38	20.8	157
296	1.66	25.2	140
375	1.89	29.2	127
444	1.89	28.9	126
519	2.16	32.9	124
667	2.28	34.4	136
743	2.26	34.1	145
811	2.26	33.9	147
883	2.24	33.5	150
940	2.20	32.9	153
999	2.15	32.1	157
1058	2.08	31.1	161
1123	2.04	30.4	167
1174	1.94	28.9	173
1221	1.84	27.8	181
1285	1.71	25.9	194
1301	1.70	25.7	196
1346	1.64	24.3	206
1401	1.56	23.1	214
1552	1.38	20.8	235
1703	1.31	20.1	245
1851	1.28	19.3	252
1992	1.27	19.2	253
2140	1.28	19.5	253
2290	1.28	19.5	257
2741	1.30	19.7	259
3053	1.30	20.1	259
3503	1.30	19.3	265
4056	1.14	21.3	260

Graph (A)
Phosphate Profile

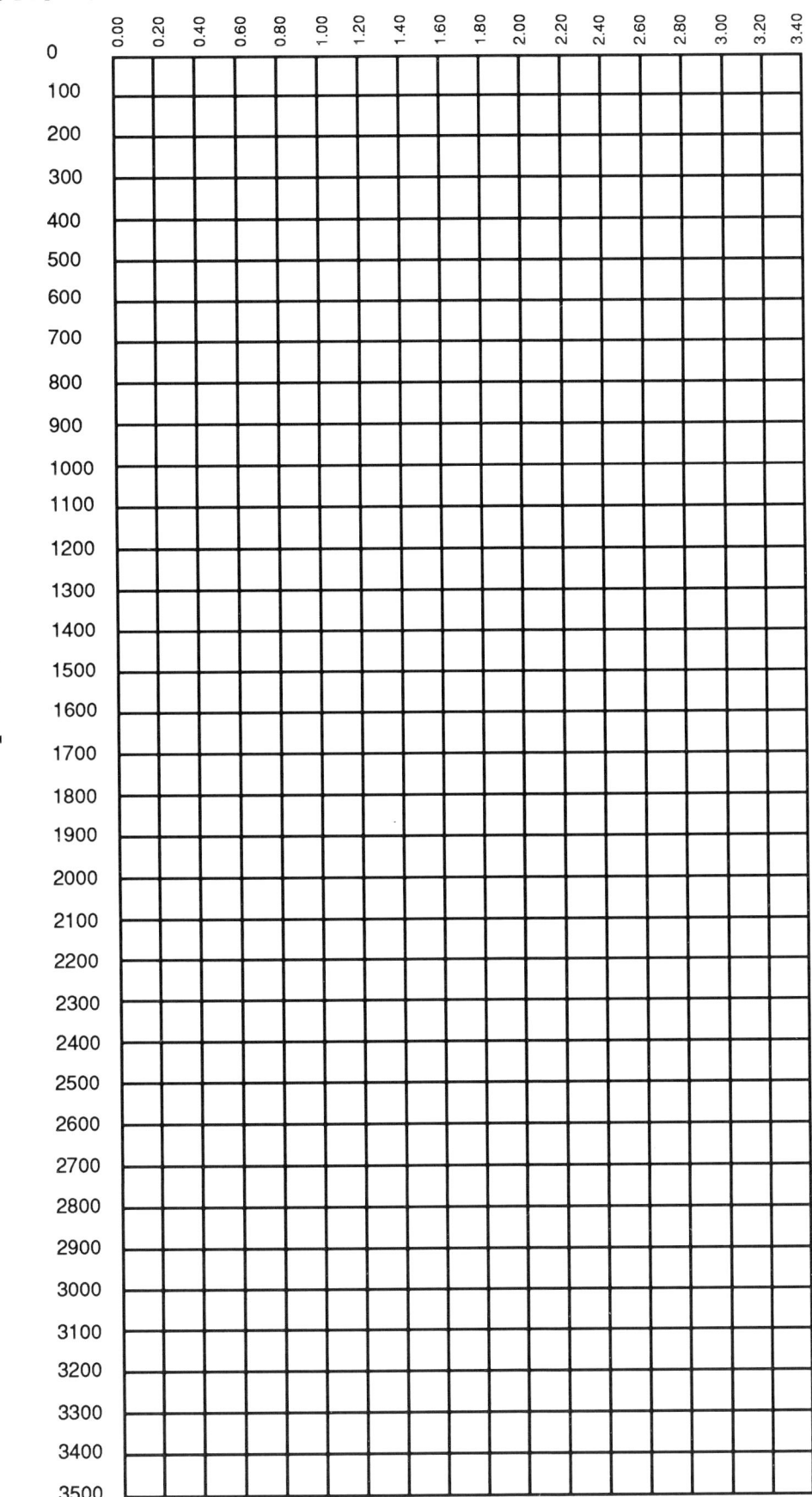

Graph (B)
Nitrate Profile

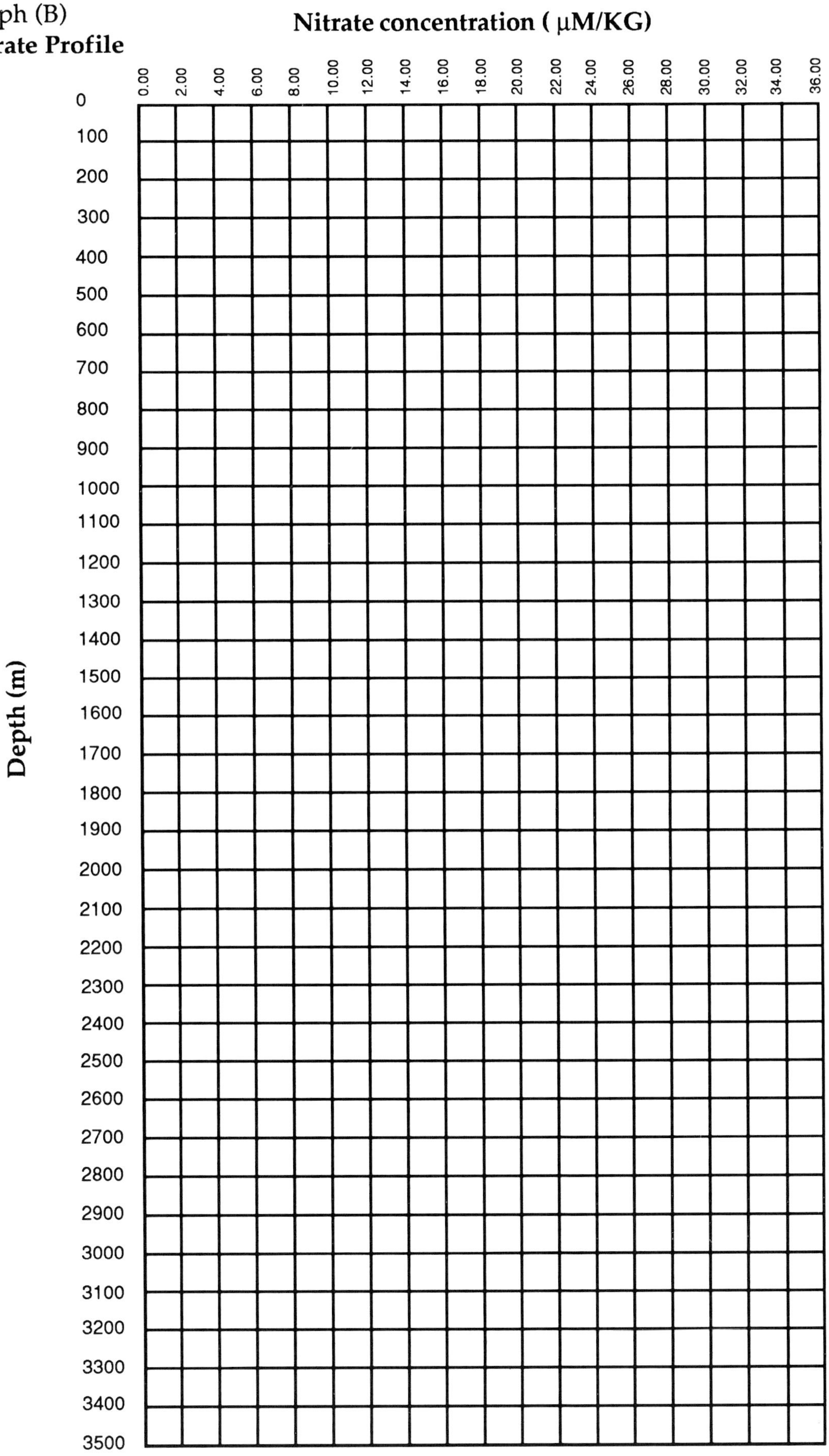

Nitrate concentration (μM/KG)

Depth (m)

Graph (C)
Oxygen Profile

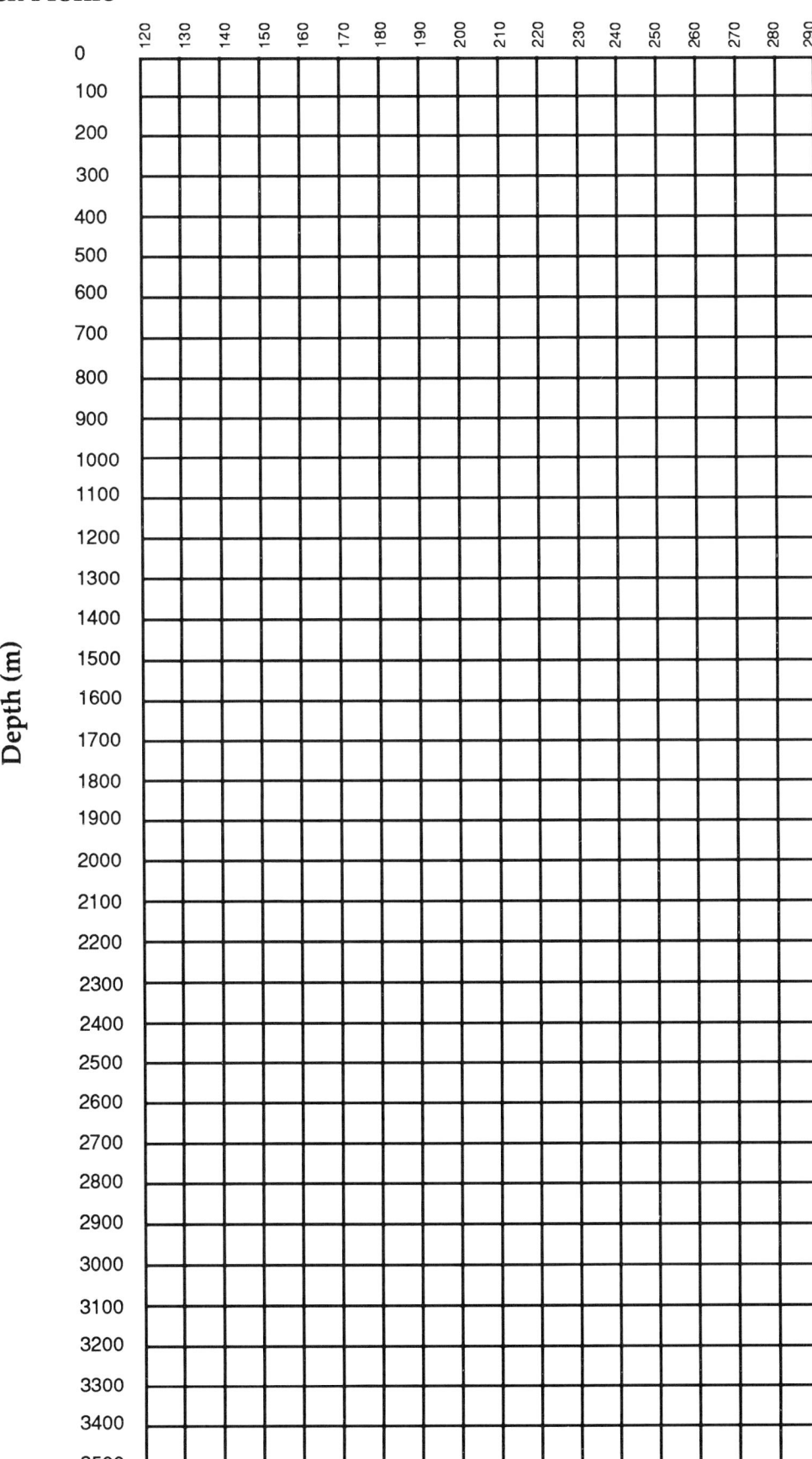

Additional Questions for Review

1. From what you know of vertical stratification and vertical nutrient distribution, where would you say high surface productivity would occur?

2. Define upwelling.

3. How would upwelling effect surface productivity in low latitude surface waters?

4. What are the important nutrients needed for marine productivity?

PLANKTON

Plankton are floating organisms. In general they are **pelagic** organisms (organisms in the water column) that are unable to swim against a current of, say, one knot for an extended period of time. Plankton tend to drift with the ocean currents, though some can perform regular vertical migrations. Pelagic animals capable of swimming are called **nekton**. Plankton may be classified in a number of ways. They can be grouped into **phytoplankton** (photosynthesizing organisms) or **zooplankton** (non-photosynthesizing organisms). They may also be classified by habitat, such as *oceanic* (offshore) or *neritic* (nearshore) plankton, *epipelagic* (surface to 200 meters) or *mesopelagic* (200 to about 1000 meters) plankton. Sometimes special terms are used for plankton occupying specific habitats. Also, some plankton, the *holoplankton*, exist as such throughout life, whereas *meroplankton* are planktic for only a portion of their life cycle, as are clams at the juvenile stage or fish at the larval stage. Finally, plankton can be classified according to size, as in the following list.

megaplankton	- larger than 2000 microns
mesoplankton	- 200 - 2000 microns
microplankton	- 20 - 200 microns
nannoplankton	- 2 - 20 microns
ultranannoplankton	- smaller than 2 microns

Phytoplankton make their own food (i.e. they are autotrophic) by photosynthesis. In order to photosynthesize, plants require sunlight, nutrients, water, and carbon dioxide. Typically, sunlight and availability of nutrients limit the growth of marine phytoplankton. Visible light penetration, relative to that of the atmosphere, decreases with depth in water, so that **irradiance** (the supply rate of radiant energy) decays exponentially with increasing depth. Light levels sufficient for photosynthesis extend downward to depths of approximately 100 meters. This depth is known as the photic zone.

The most important and abundant phytoplankters are **diatoms** (photosynthetic organisms with silica (SiO_2) shells). They typically flourish in cold, nutrient-rich water, particularly in the polar and subpolar latitudes and the inshore and shelf waters of the mid-latitudes. When light and nutrient levels are favorable, diatoms can divide (reproduce) once every 12 - 24 hours, producing a dense **bloom**. During adverse conditions (low light levels during winter months or low nutrients) some diatoms remain dormant until conditions favorable to reproduction reappear.

Dinoflagellates are another important marine plant. Though they are found throughout the oceans, they favor warm, tropical waters. Under certain conditions their numbers can exceed those of diatoms. Some species of dinoflagellates (zooxanthellae) live inside the cells of corals and foraminifera (see diagrams). The coral animal and the dinoflagellate benefit mutually from their association. This biological association is known as **mutualism** or symbiosis.

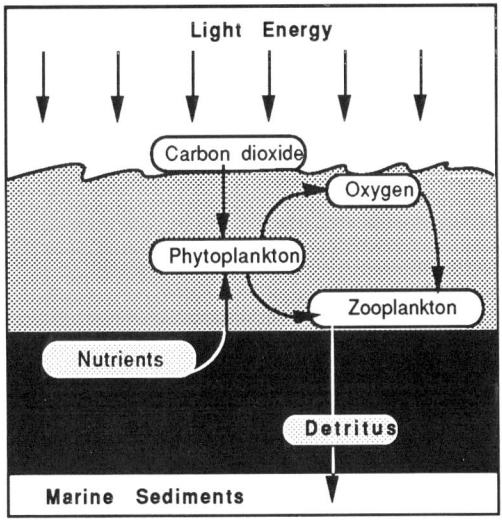

FIGURE 12-1. Oceanographic processes important to global biogeochemical cycles.

Zooplankton cannot make their own food (i.e. they are heterotrophic) and thus graze on phytoplankton for their nutrients. **Copepods** (crustaceans) and **foraminifera** (single celled organisms that secrete a shell composed of calcium carbonate) are two examples.

Definitions

Autotrophs. Plants and bacteria that synthesize food from inorganic nutrients.

Detritus. Inorganic or organic debris.

Heterotrophs. Animals and bacteria that require prefabricated food for sustenance.

Nekton. Animals that can swim independently of current flow.

Phytoplankton. Plant plankton. The primary producers of the oceans.

Exercise

Refer to the color images on pages 136 through 139.

The following are color satellite images of the annual phytoplankton growth cycle in the North Atlantic. Within the ocean basin shading represents levels of phytoplankton growth. Study the images and then answer the following questions.

> **Use the Color Code bar on page 136 to determine productivity levels.**

1. Where in the North Atlantic does productivity increase as the year progresses from January to December?

2. What causes this bloom?

3. What is the movement of the bloom throughout the year? (How does it propagate?)

4. On the average, what regions contain the greatest productivity all year long? Explain why this occurs?

4. Notice the region near the Amazon river. During what time of the year is productivity increasing? Why?

5. Why is there very little productivity in tropical regions?

To answer the following questions, refer to the Global Biosphere satellite image on page 140.

> Use the Color Code bar on page 136 to determine productivity.

6. Generally speaking, where are the zones of highest productivity?

7. Explain why there is a band of high productivity along the equator in the Pacific Ocean.

8. What can you say about the productivity within oceanic gyres?

9. Explain why productivity appears to be higher on the western coasts of South America and Africa.

Exercise

Examine the plankton samples provided under the microscope. For each sample answer the following questions. Use the charts provided as a guide to identification.

1. Identify as many organisms as you can in the samples provided.

2. What is the composition of each of these plankters? (e.g. $CaCO_3$, SiO_2, cellulose, chitin)

3. What are the dominant plankton types represented? (phtyo- vs. zooplankton and type)

4. Do these organisms tend to live in nutrient-rich or relatively nutrient-poor waters?

5. During a bloom, would you expect all of these organisms to appear at once or would there be a sequence to their development? In either case **explain** your reasoning.

6. If the waters in which these samples were collected were to become depleted in silica what would you expect to occur?

PHYTOPLANKTON

DIATOMS (Contain skeletons composed of silica)

Solitary Centric Diatoms 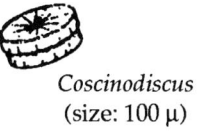 *Coscinodiscus* (size: 100 μ)

Colonial Centric Diatoms 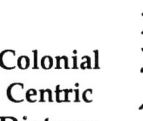 *Chaetoceros* (size: 100

Pennate Diatoms  *Nitzschia* (size: 100 μ)

Rhizosolenia (size: 100 μ)

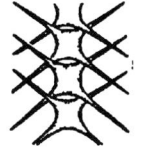 *Eucampis* (size: 100 μ)

DINOFLAGELLATES (Contain skeletons composed of organic matter)

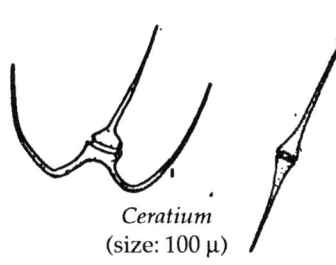
Ceratium (size: 100 μ)

 *Gonyaulax* (size: 50 μ)

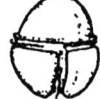 *Gymnodinium* (size: 50 μ)

 Peridinium (size: 50 μ)

COCCOLITHOPHORES (Covered by calcareous plates)
(individual plates are known as coccoliths)

Coccolithus
(note: organism is covered by coccoliths)
(size: 25 μ)

 Individual coccoliths (size: 5 μ)

SILICOFLAGELLATES (Contain skeletons composed of silica)

Dictyocha (size: 50 μ)

FLAGELLATES (Contain no skeletons)

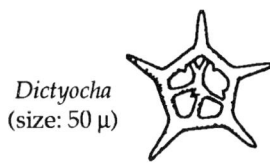
(size: 5-10 μ)

ZOOPLANKTON

FORAMINIFERA
(Skeletons composed of calcium carbonate)

Globigerina
(size: 100 μ)

RADIOLARIANS
(Contain skeletons composed of silica)

Lamprocycias
(size: 100 μ)

Spumellarians
(size: 100 μ)

COELENTRATES (Jelly fish and other forms)

Turritopsis

Eutima

(size varies from a few centimeters to a meter)

PTEROPODS (Contain shells composed of aragonite)

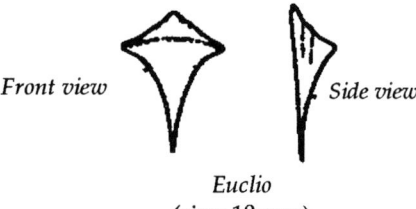

Front view *Side view*

Euclio
(size: 10 mm)

Limacina
(size: 5 mm)

COPEPODS

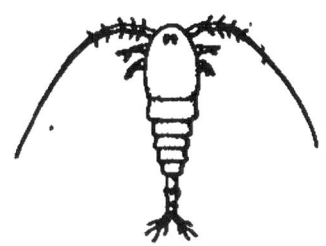

Calanus
Calanoid copepod
(size: 1-5 mm)

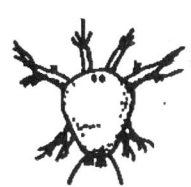

Nauplius larva
(larval stage of copepods and
other crustaceans)
(size: 1 mm)

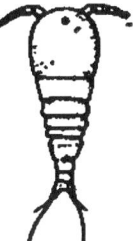

Halicyclops
Cyclopoid copepod
(size: 1-5 mm)

CRUSTACEANS

Euphausiids

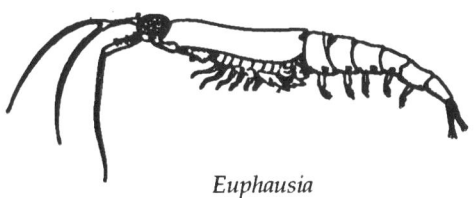

Euphausia
(size: 1-6 cm)

MEROPLANKTON OF BENTHIC INVERTEBRATES

Clam larva
(size: 1 mm)

Veliger larva
(snail)
(size: 2 mm)

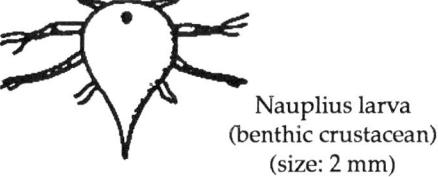
Nauplius larva
(benthic crustacean)
(size: 2 mm)

DEEP-SEA VENT COMMUNITIES

Deep-sea vent communities are found on the ocean floor in the vicinity of mid-oceanic ridges. Unknown prior to 1977, these fascinating assemblages of organisms were discovered by scientists using the submersible craft "Alvin" to make deep-sea dives as deep as 2700 m. The organisms that are a part of this ecosystem actually need the vents in the ridge to survive, so it is not by chance that we find these organisms there.

Recall that the mid-ocean ridge is the spreading center of the ocean floor, and that the center of the ridge is volcanic, ejecting molten lava and gases. One of the gases released is a reduced form of **sulfur**. Life around the vents is dependent upon this supply of sulfur.

At the base of this community's food chain is bacteria. Since such producers form the base of a food chain, they must be autotrophic (produce their own food). The producers that we are already familiar with, the phytoplankton, are photosynthetic, using light to produce food energy. However, at the bottom of the sea there is no light for photosynthesis to take place. Therefore, another reaction, called **chemosynthesis**, occurs in its place. The bacteria producers are chemosynthetic, using a chemical (chemo-) to manufacture food (-synthesis). This chemical is sulfur. The chemical reaction that occurs utilizes available oxygen and carbon dioxide (both dissolved in seawater) and hydrogen sulfide (emitted from the vents) to produce organic matter:

$$H_2O + O_2 + CO_2 + H_2S \rightarrow CH_2O + SO_4^{2-} + 2H^+$$

(CH_2O is the generic form of organic matter.)

Given the great depth of these communities in the ocean, we might expect the habitat to be very cold. However, water temperature around the vent is 8-16 degrees centigrade, as compared to 2 degrees at other areas of the same depth. This is due to the presence of the active vents. Thus the entire vent community is limited to an area that is only a few meters from the vent.

The organisms inhabiting the vent habitat are found nowhere else in the world. The world's largest clam, *Calyptogena magnifica*, lives in the deep-sea vent community. The oldest, *Calyptogena*, is estimated to be about 30 years old, therefore it is inferred that individual communities only live on the order of decades. Given the short life spans of vent organisms, we know that there must be tremendous pressure for them to quickly and efficiently reproduce.

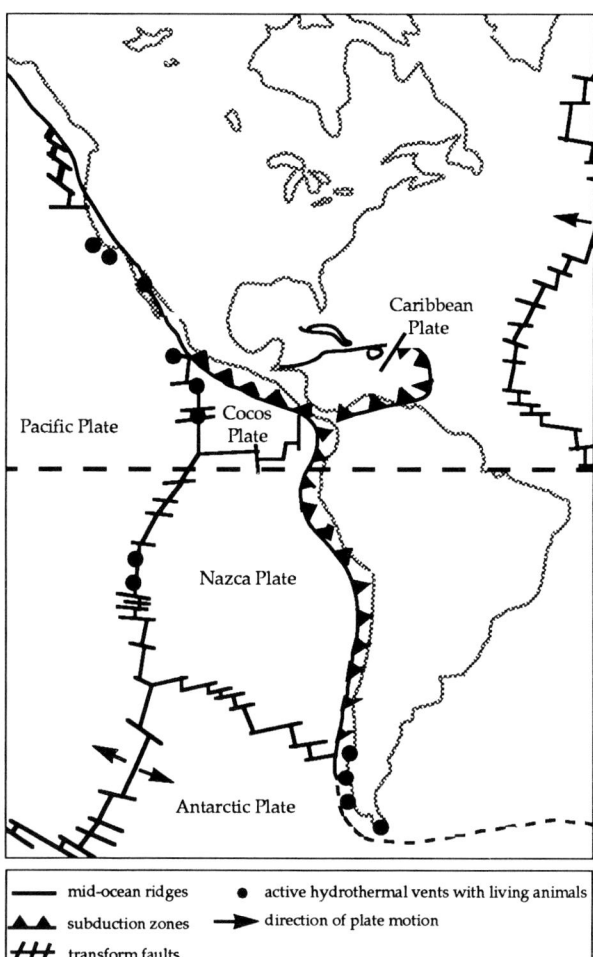

FIGURE 11-1. Location of some deep sea vent communities.

Definitions

Chemosynthesis. The name for the reaction employed by deep-sea bacteria which use sulfur to produce food.

Photosynthesis. The process by which plants convert carbon dioxide and water into carbohydrates, using sunlight as a source of energy,

Sulfur. A chemical element emitted by volcanoes that sustains life at deep-sea vent communities.

Additional Questions for Review

1. Where on the globe are the scientists in the film studying? Do you think that vent communities can only be found in the ocean that they are exploring? Why or why not?

2. Think of reasons why it would be difficult to keep deep-sea vent organisms alive at the surface.

3. The same species of deep-sea vent organisms are found distributed (patchy) along all mid-ocean ridges. How might they spread from one site to another?

SALT MARSH COMMUNITIES

The salt marsh is a rigorous environment in which few plants or animals live. Salt marshes are **low-energy, tidally-dominated** areas that typically have high soil salinities. The soils, or **substrates**, are usually peats. fine-grained silts or clays that retain water well. A low-energy, tide-dominated environment is one which is rarely or never impacted by ocean waves, but has its lowest areas flooded twice daily, or **semi-diurnally** by high tides.

Despite the harsh conditions, there are some organisms that have adapted to the salt marsh habitat. Plants living on the marsh are called **halophytes**, and they form the base of the **trophic pyramid** and the salt marsh **food chain.** Trophic is a word which biologists use to refer to feeding.

A simple food chain

Food chains show the flow of energy in an ecosystem, or simply, who eats what. The most basic outline for a food chain is as follows:

producers → primary → secondary → tertiary
            consumers   consumers   consumers...

Producers are generally the plants of the community. They produce all the food that is initially available, and therefore are responsible for sustaining all levels of the food chain, either directly or indirectly. They are **autotrophic**, meaning they produce all their own food. In contrast, all other members of the food chain, which need to eat to obtain nutrients necessary for survival, are called **heterotrophic**, which means they need to rely on sources outside their bodies.

Consumers are the heterotrophs. *Primary consumers* are those which directly eat the producers, *secondary consumers* eat the primary consumers, and so on until the top of the food chain, which consists of animals which are not eaten by anything else in the community. Sometimes these animals are referred to as the **top carnivores.**

The trophic pyramid

Producers are the most abundant part of the food chain. Each successively higher trophic level has fewer organisms than the one below it. This structure may be thought of as a pyramid, with producers forming the base:

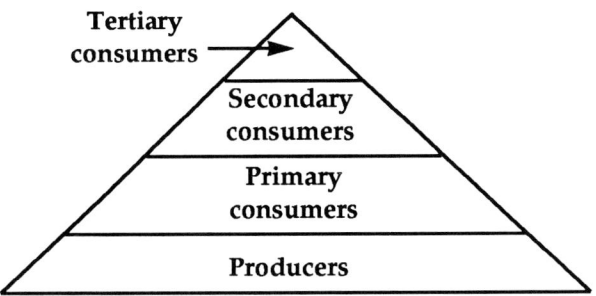

The food web

A food chain is a simplified way of looking at feeding structures in an ecosystem. In reality, organisms tend to eat more than one type of food. This fact may be better represented by a **food web**, which shows the more complex interactions within a feeding community. For example, secondary consumers eat the primary consumers, but they may also eat producers directly. In the end, everything dies and is decomposed by bacteria collectively referred to as the **decomposers.** In a trophic web, energy flow ultimately ends with the decomposers. Also, The terms primary and secondary consumer are somewhat vague, and it is useful to break them down into categories which we can better understand.

Animals which only eat plants, or producers, are called **herbivores**. **Herbivores** are primary consumers. Sometimes marsh plants are weathered into tiny pieces, which are called **detritus**, and those organisms which eat this are specifically called **detritus feeders. Carnivores** eat other animals, and **omnivores** eat both plant and animal food. Figure 1 shows an example of a simple estuarine food web.

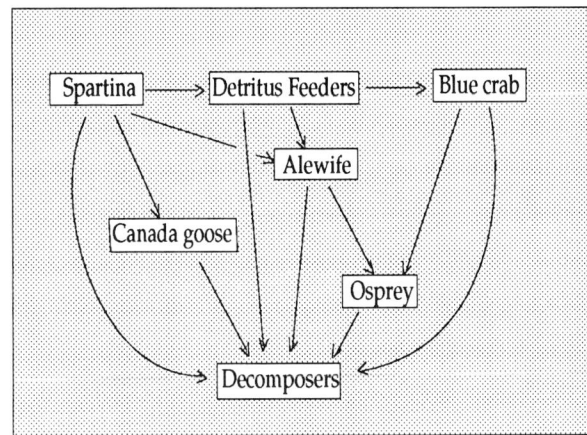

FIGURE 12-1. Food web.

Organisms of the Chesapeake Bay estuary and salt marsh

This is a list of organisms of the Chesapeake Bay and their trophic categories.

Producers
    Spartina alterniflora - cordgrass
    Spartina patens - salt marsh hay
    Juncus gerardi - spike rush
    Phragmites australis - common reed
    Phytoplankton
    Zostera marina - eelgrass

Detritus feeders
    Uca pugnax - fiddler crab
    Shrimp
    Various worms

Herbivores
    Branta canadensis - Canada goose
    Aythya valisineria - Canvasback duck

Omnivores
    Callinectes sapidus - blue crab - will eat almost anything on the marsh, living or dead, including other crabs of its species.
    Anas rubripes - black duck - eats small fish and plants.
    Alosa pseudoharengus - alewife (fish) - eat algae and zooplankton.
    Alosa aestivalis - blueback herring - eat zoo- and phytoplankton.

Carnivores
    Haliaeetus leucocephalus - bald eagle
    Falco peregrinus - Peregrine falcon
    Ardea herodias - Great blue heron
    Pandion haliaetus - Osprey

Energy transfer within a food web

When an organism eats, or consumes nutrition, an energy transfer has taken place from one trophic level to another. Food has an energy value, often represented in kilocalories, or kcal. For example, a patch of grass that has an area of 5 square centimeters has a certain number of kcal. If a duck eats all of that grass, it gets energy by consuming kilocalories. This progresses through the chain, and the result is a net energy transfer to the top of the food chain.

In a trophic system, not all of the energy from one trophic level is passed on to the next level. Some of it is left behind to reproduce or decay. **Ecological efficiency** is a measure of the percentage of the energy passed from one level to the next. It is represented by:

$$\text{Ecological efficiency} = \frac{\text{amount of energy in a trophic level that is consumed}}{\text{total amount of energy in the trophic level}}$$

It is generally accepted by biologists that the ecological efficiency between two trophic levels is generally between 10 and 20%. Progression through the food chain reflects a net decrease in original energy from the producers with each successive level.

Definitions

Autotroph. An organism that produces all its required nutrition within itself, e.g. a producer.

Consumer. An organism that must obtain nutrition from sources outside its body.

Decomposer. Bacteria which returns dead matter to its original organic state.

Detritus. Broken off bits of plants which become weathered.

Ecological efficiency. The percentage of energy in a trophic level that is transferred directly to a higher level.

Food chain. A simple representation of the flow of energy in a trophic community.

Halophyte. A plant which grows in saline conditions and nowhere else.

Herbivore. An animal that eats only plants.

Heterotroph. An organism that needs to eat to live, e.g. a consumer.

Omnivore. An animal which eats both plants and other animals.

Producer. An autotrophic and/or photosynthetic organism at the base of the food chain or web.

Top carnivore. The animal at the top of the food web, chain, or pyramid which is not eaten by anything else in the community except decomposers after death.

Trophic. A term which refers to feeding in a community.

PENN STATE BOOKSTORE ON CAMPUS

```
128 CASH-1            7435 0001 005

9780878720063 NEW
SEA AROUND US LA    MDS 1N     19.50
                    TOTAL      19.50

CASH TENDER                    20.00
      CHANGE                     .50
    WE ARE...PENN STATE!!!
```

9/30/94 2:46 PM

PENN STATE BOOKSTORE ON CAMPUS

COO LOGO ZIP   1-8341 BS
                    128 CASH 1

9780757200035 NEW
SEA AROUND US 1A   19.50
                ADD A1 50   10.4L
                            TOTAL 19.50

CASH TENDER   20.00
                CHANGE   .50

WE ARE... PENN STATE!!

9780757200035

Exercise

Using the list of organisms of the Chesapeake Bay, construct your own simple food web. Do NOT attempt to do this using the boldface categories; rather, select two species from each category and construct your web based on what each might eat specifically. **Arrows should point from the food to who is eating it, indicating the direction in which the energy is flowing.**

Additional Questions for Review

1. (a) The Chesapeake Bay watershed consists of 135,000 acres of marsh. There are 4044 square meters in an acre. How many square meters of marsh are in the watershed?

   (b) The net productivity of the Chesapeake marsh producers is 6850 kcal/m^2/year. How many kilocalories of energy are produced in the marsh each year?

   (c) If 5.609×10^{11} kcal are consumed from the marsh by primary consumers per year, what is the ecological efficiency between the producers and the first trophic level?

2. Construct another food web, using different organisms than the ones used in class.

3. Which do you think would have a higher ecological efficiency and **why**: A trophic system with many trophic levels or one with few levels?

4. If 5.609×10^{11} kcal are consumed from the marsh by primary consumers per year, what is the ecologic efficiency between the producers and the first trophic level?

5. Which do you think would have a higher ecological efficiency and **why**: A trophic system with many trophic levels or one with few levels?

Answers to Additional Questions

Navigation

1. 24,903 statute miles; 21,655 nautical miles
2. 1,037.6 statute miles/hour or 902.29 nautical miles/hour
3. slower; a shorter distance is traveled in the same amount of time
4. 32°40′
5. 1/4°/minute or .25°/minute; .004°/second or 1/4′/second
6. 20°10′ 15″ N, 29°05′ 12″ W
7. 8 am @ point A; 8 pm @ point B
8. 20 nautical miles; 23 statute miles
9. 45°W; cannot determine latitude
10. (a) 1,200 nautical miles; 1,380 statute miles
    (b) 16°35′
    (c) 1 hour

Bathymetry

1. 3,675 meters; 12,000 feet
2. 200x
3. (a) hill-like feature
   (b) steeper
   (c) 500 meters
4. 6 seconds

Sea-Floor Spreading

1. The sea-floor comprising the mid-ocean ridges are hotter, and therefore less dense than the surrounding sea-floor. This less dense material will then rise above the surrounding area due to isostacy and upwelling of magma.
2. Continental crust is thicker and less dense than oceanic crust.
4. shearing; compression; tensile
7. Subduction creates a tremendous amount of heat due to friction and melting of oceanic crust deep in the mantle.

Wave

2. 250 cm or greater
3. 6 meters

Salinity

1. Freshwater merges very quickly with saltwater.
2. Freshwater is on top because of lower density. Saltwater sinks below freshwater forming a salt water wedge.
3. As the sea moves out with the tide, fresh water fills the estuary. The reverse is true during high tide. Organisms that live in this environment must be tolerant of rapid salinity changes.

Ocean Stratification

2. (a) 0.044 °/meter
   (b) 0.032 °/meter

Ocean Nutrient Dynamics

2. The slow, upward transport of water to the surface from depth.
3. Upwelling would return nutrients from deeper waters to the surface and allow for increased productivity.

Salt Marsh Community

1. (a) 5.46×10^8
   (b) 3.74×10^{12}
   (c) 15%
3. Few -- 10% per level transferred. More levels, more energy lost.

References

Figures 3-2, 6-1, 7-1, 8-2: modified from Pinet, P.R., 1992, **Oceanography: An Introduction to the Planet Oceanus**, West Publishing Company.

Figure 11-1: modified from Nybakken, J.W., 1988, **Marine Biology: An Ecological Approach**, 3rd edition, New York, Harper and Row.

Figure 8-1: modified from **GEOSECS Atlantic Expedition**, Volume 2, National Science Foundation, Washington, D.C.

Color images on pages 136 through 139 from: NSF/NASA-sponsored US Global Ocean Flux Study Office, Woods Hole Oceanographic Institution, Woods Hole, Ma.

Photo #1

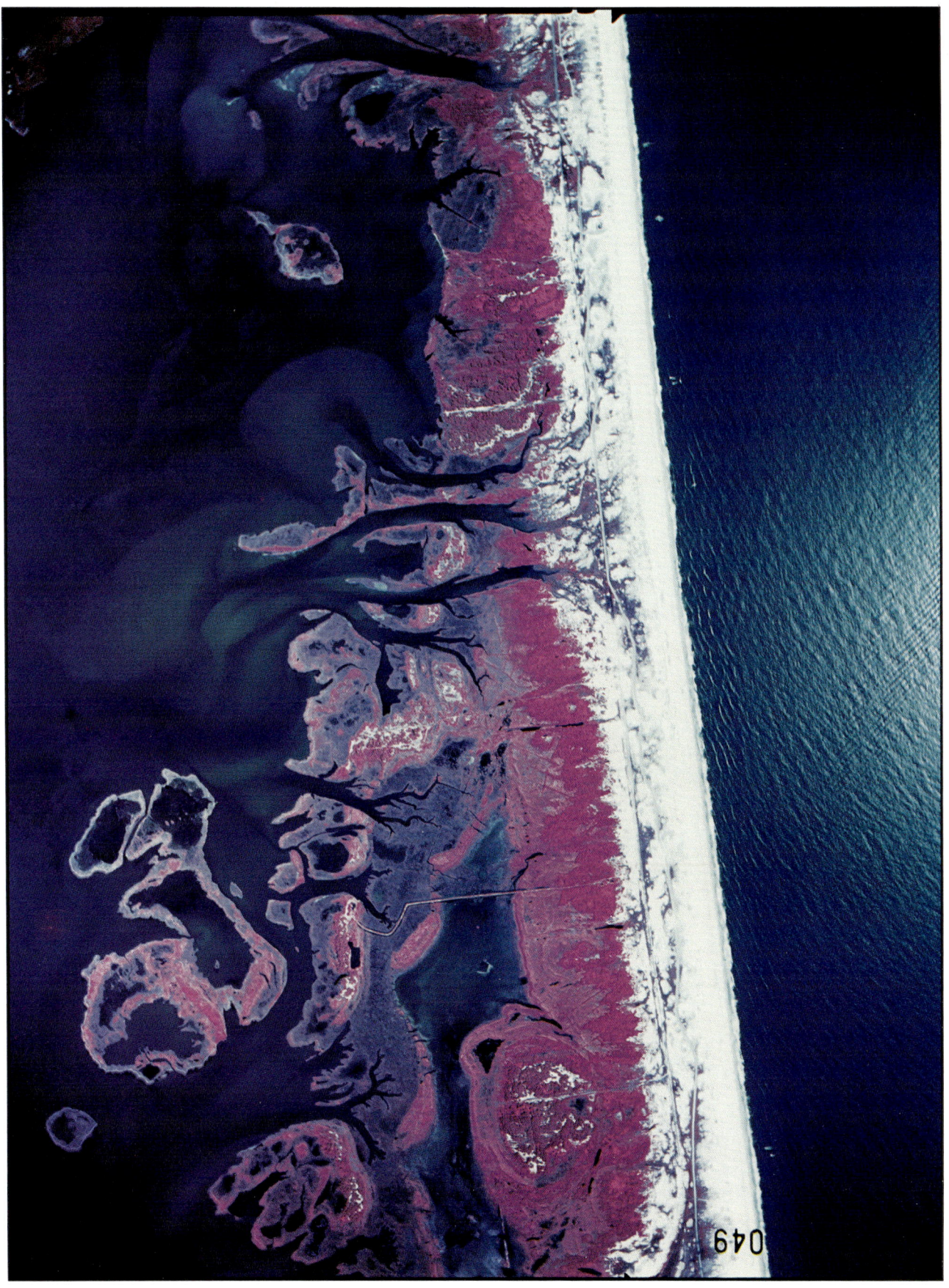

Photo #2

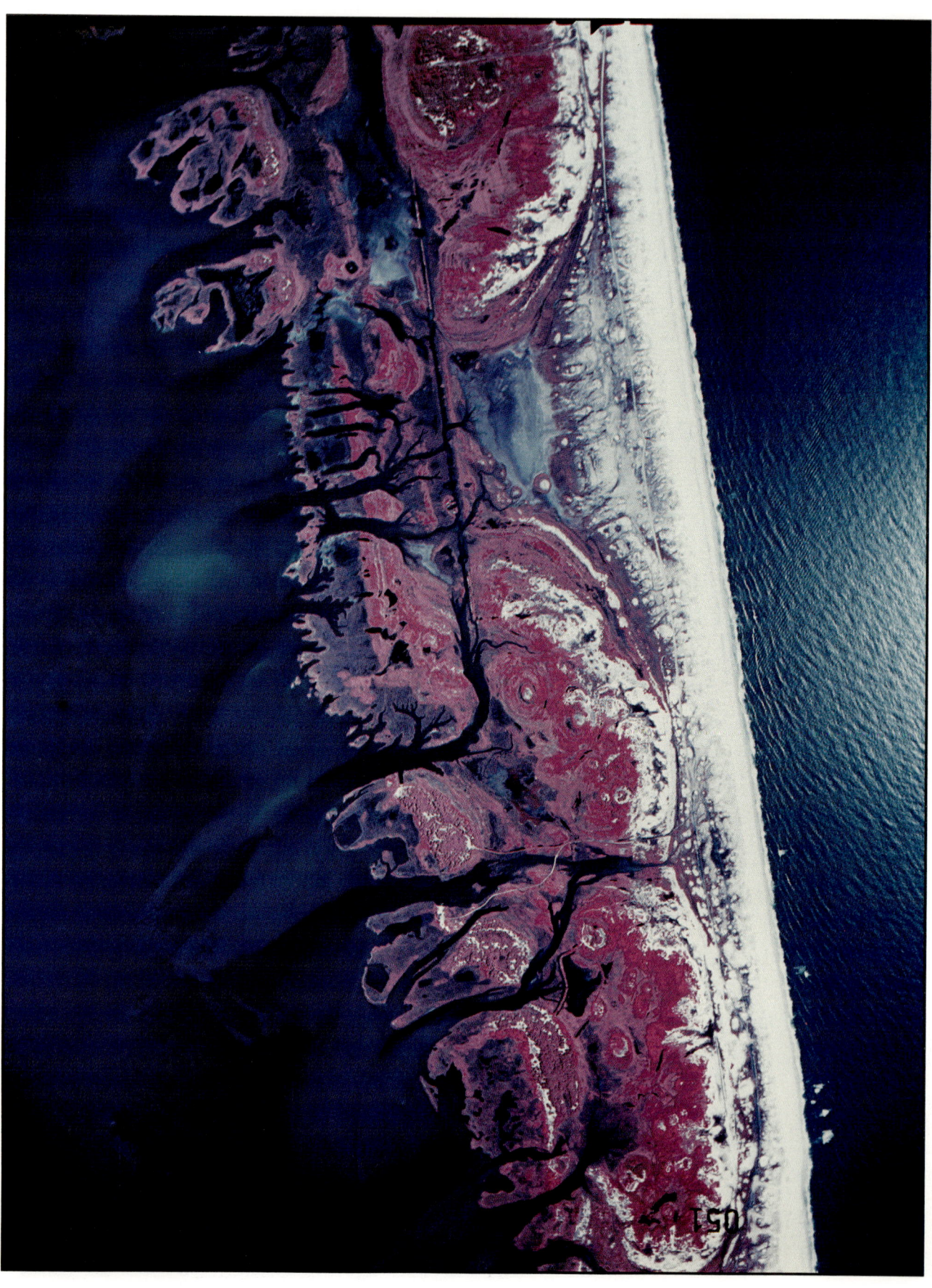

Photo #3

Photo #4

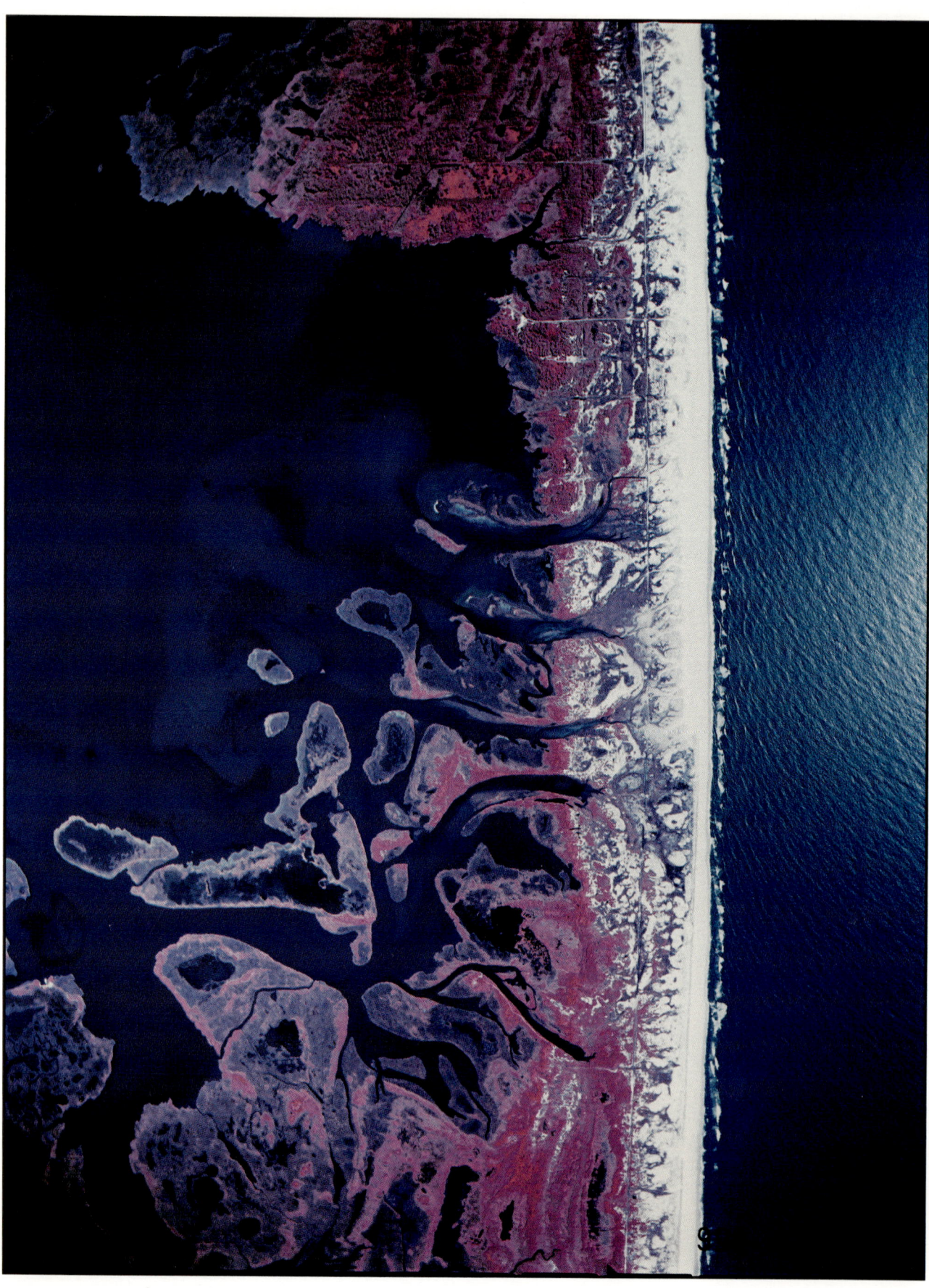

Photo #5

Photo #6

Photo #7

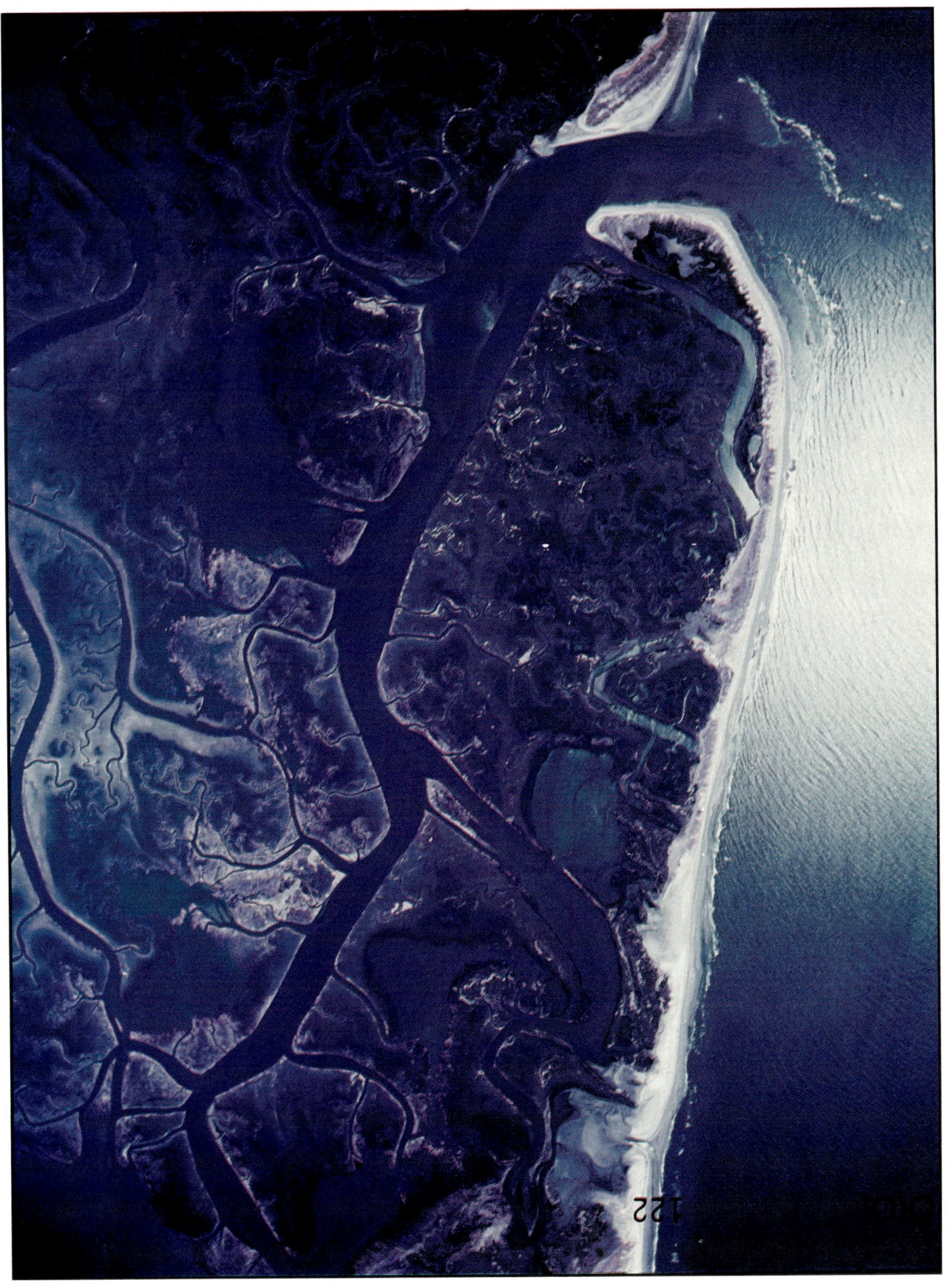

Photo #8

Photo #9

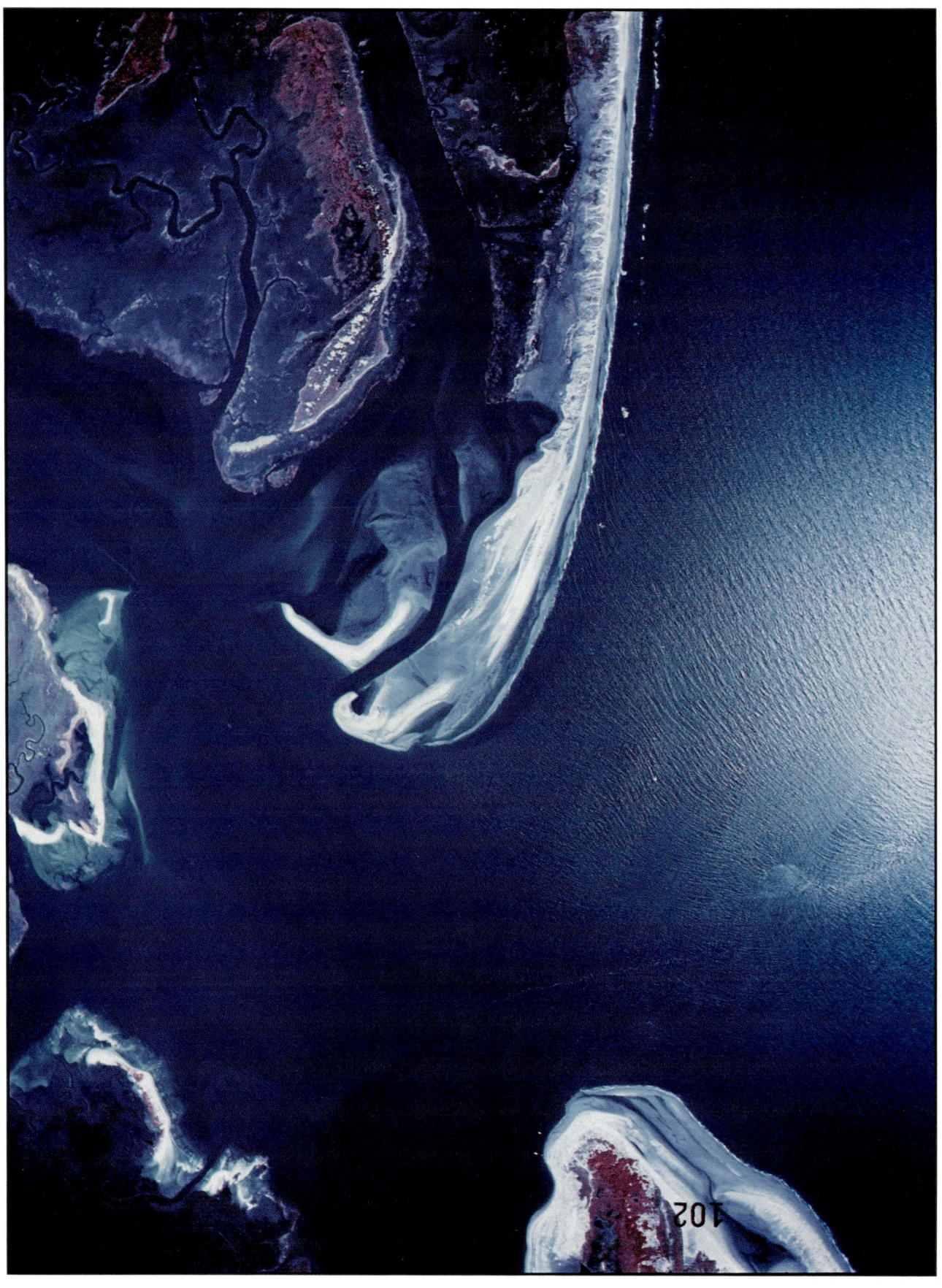

Photo #10

Photo #11

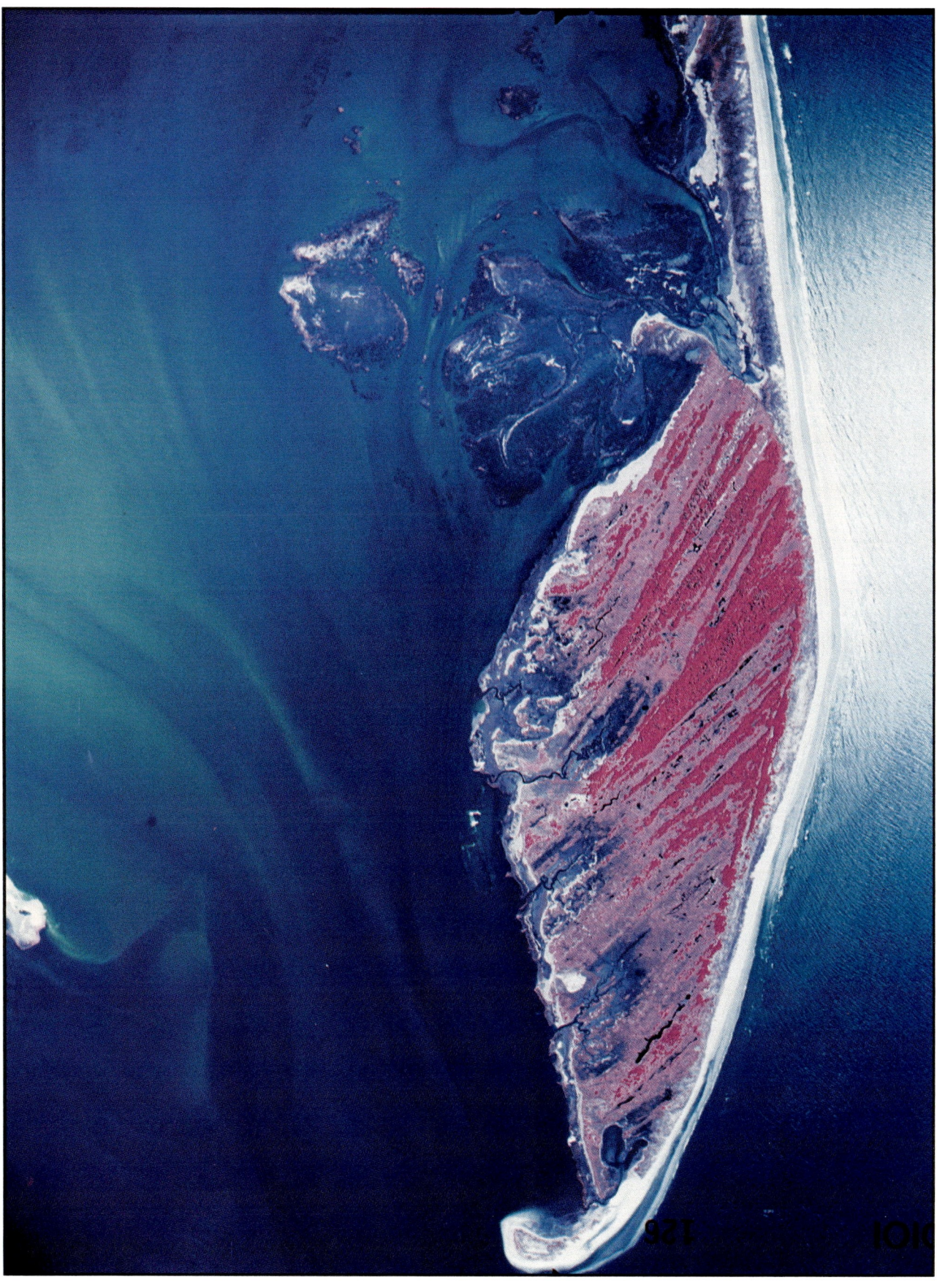

North Atlantic Phytoplankton Productivity

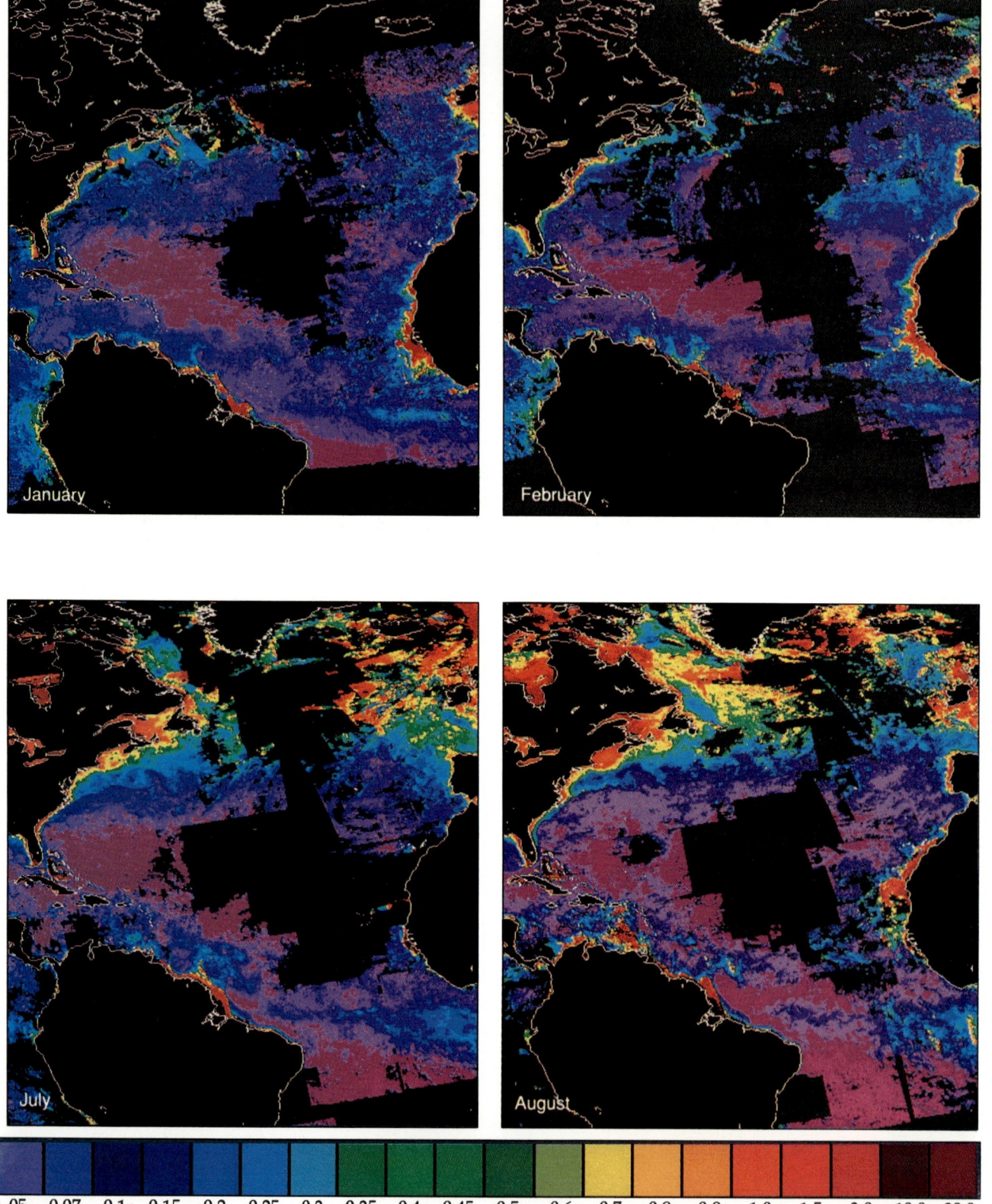

Color Code: represents chlorophyll concentration (mg/m^3)

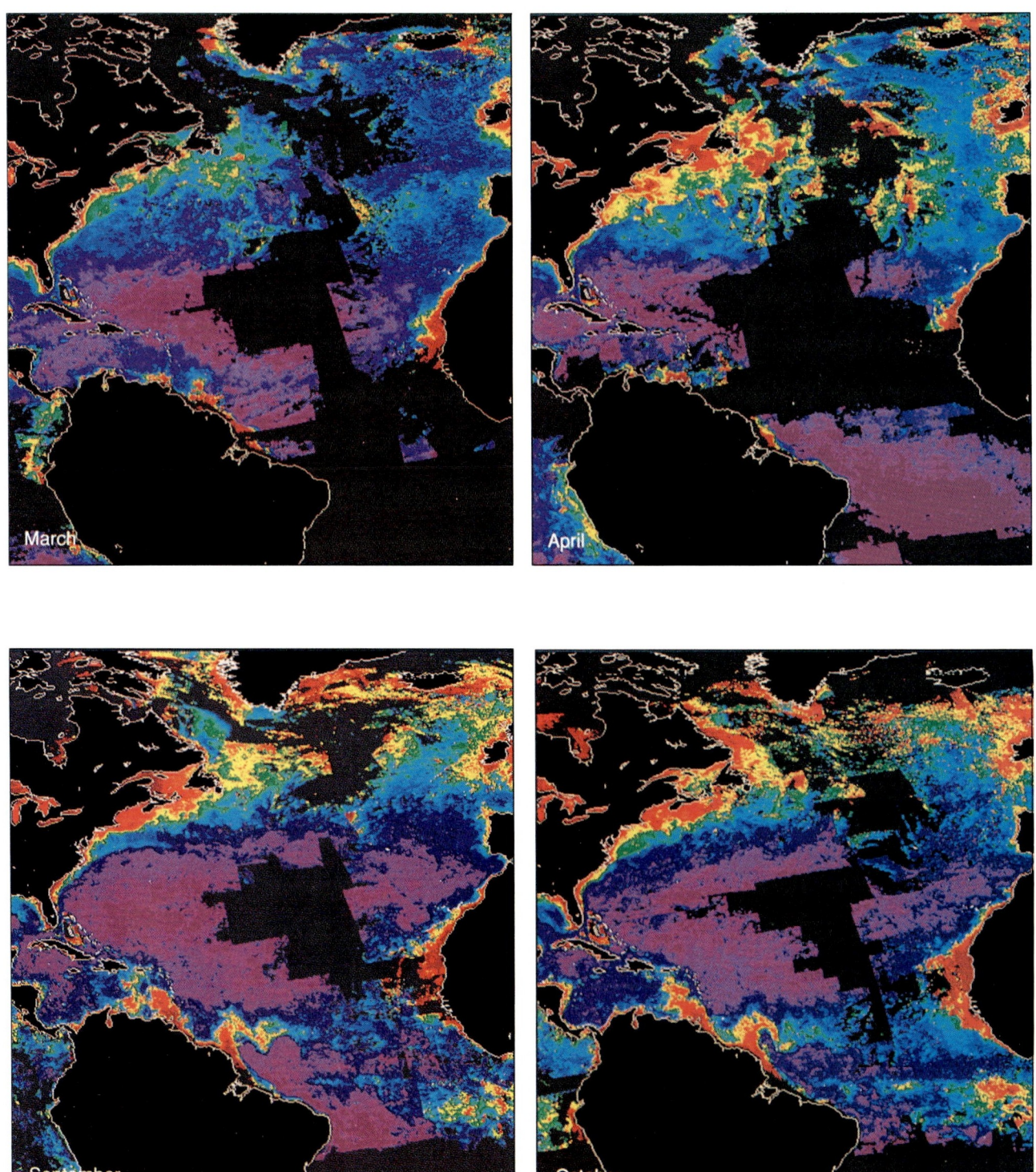

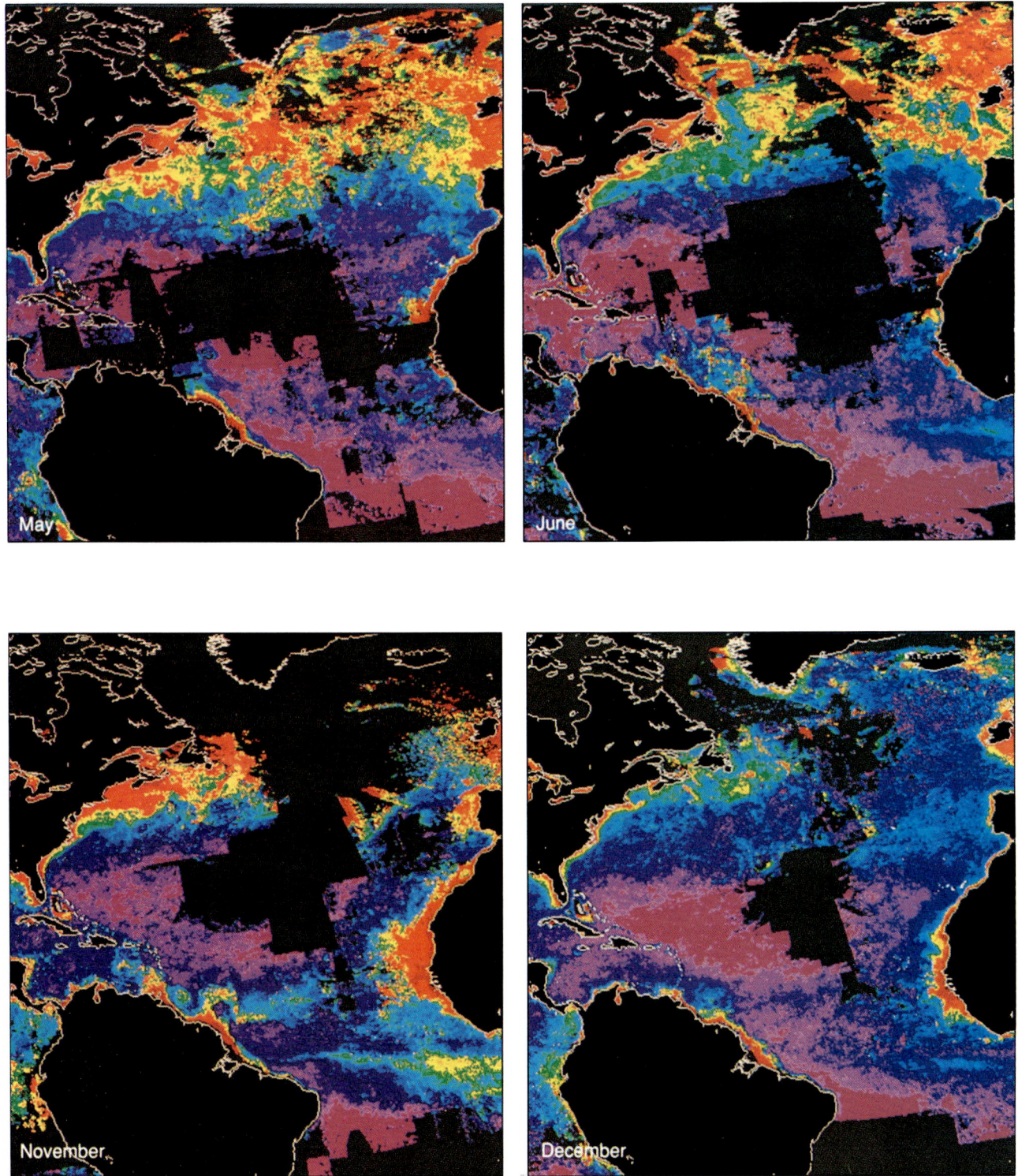

Global Biosphere

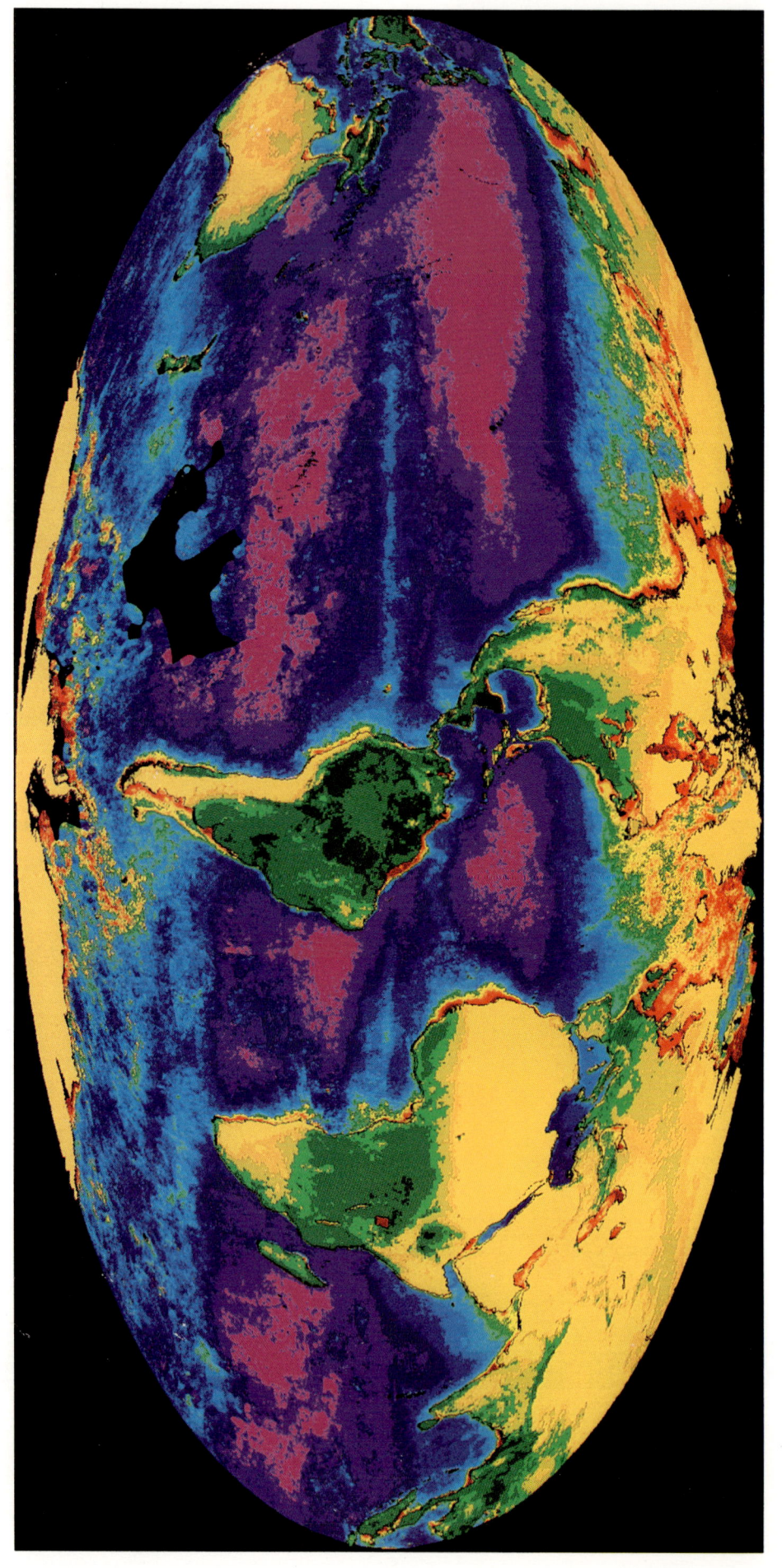